Experimental Brain Research Series 19

Walter K. Schwerdtfeger Peter Germroth (Eds.)

The Forebrain in Nonmammals

New Aspects of Structure and Development

With 51 Figures and 6 Tables

Springer-Verlag Berlin Heidelberg New York
London Paris Tokyo Hong Kong Barcelona

Priv.-Doz. Dr. Walter K. Schwerdtfeger
Paul-Ehrlich-Institut
Paul-Ehrlich-Str. 51, 6070 Langen
Federal Republic of Germany

Peter Germroth
Max-Planck-Institut für Hirnforschung
Deutschordenstr. 46, 6000 Frankfurt/M. 71
Federal Republic of Germany

ISBN-13:978-3-642-75369-5 e-ISBN-13:978-3-642-75367-1
DOI: 10.1007/978-3-642-75367-1

Library of Congress Cataloging-in-Publication Data
The Forebrain in Nonmammals / Walter K. Schwerdtfeger, Peter Germroth (eds.).
p. cm. -- (Experimental brain research series; 19)
Based on the contributions presented at a symposium held Sept. 2-3, 1988 as a satellite to the Annual Meeting of the European Neuroscience Association at the University of Zürich-Irchel.
ISBN-13:978-3-642-75369-5 (U.S.: alk. paper)
1. Prosencephalon--Congresses. 2. Vertebrates--Physiology--Congresses.
3. Physiology, Comparative--Congresses.
I. Schwerdtfeger, W. K. (Walter K.), 1949- . II. Germroth, Peter, 1958- . III. European Neuroscience Congresses (11th: 1988: University of Zürich-Irchel) IV. Series.
[DNLM: 1. Telencephalon--anatomy & histology--congresses.
2. Telencephalon--growth & development--congresses. 3. Vertebrates--congresses.
W1 EX4855 v. 19 / QL 937 F714 1988]
QP376.F66 1990 596'.0188--dc20 DNLM/DLC for Library of Congress 90-9733

Softcover reprint of the hardcover 1st edition 1990

2125/3140-543210 - Printed on acid-free paper

Preface

This volume is based on the contributions presented at the symposium "Comparative Aspects of the Structure and Development of the Forebrain in Lower Vertebrates", held on September 2-3, 1988, as a Satellite to the Annual Meeting of the European Neuroscience Association at the University of Zürich-Irchel. Within the two days the symposium lasted, the main topic, of course, had to be covered in an exemplary rather than a comprehensive way. However, the comparative interpretation of results revealed many similarities of forebrain organization in the wide spectrum of species considered. Besides a great deal of new data on forebrain structure and development, this book contains attempts based on various new methodological approaches to find homologies between brain structures of nonmammalian vertebrates and mammals.

Although we are still far from having unequivocal solutions for the latter problem, the contributions to this volume illustrate impressively that comparative neuroscientists have better chances than ever of gaining new insights since they now have at their disposal a still increasing number of refined anatomical and physiological techniques such as immunocytochemistry, tract tracing of injected markers, and intracellular stimulation and recording. A prominent contribution from the rapidly evolving field of molecular biology can be anticipated. A lot of stock taking will have to be done; yet, the bulk of studies using modern methods have been dealing with the mammalian brain. Nevertheless, the international meetings of the past few years indicate that concomitantly with the methodological advances, interest in comparative neurobiology has been growing, a tendency that hopefully will continue in the future.

The editors wish to thank all the participants for their readiness to present their results and opinions and for their cooperation while this book was being compiled. It is with gratitude that we mention the support rendered by the Max-Planck-Gesellschaft, Amersham Buchler GmbH & Co KG, Plano (W. Plannet GmbH) and the

University of Zürich-Irchel, which generously provided the necessary facilities. Last but not least, we thank the staff of Springer-Verlag for their help during publication of this book.

Frankfurt/Main, July 1989

W. K. Schwerdtfeger
P. Germroth

Contents

List of Contributors*

* The address of the authors is given on the first page of each contribution

[1] Page, on which contribution commences

Secondary Olfactory Projections in *Gnathonemus petersii* (Mormyridae): A Comparative Perspective

D.J. Rooney* and T. Szabo
Department de Neurophysiologie Sensorielle , Laboratoire de Physiologie Nerveuse C.N.R.S., F-91198 Gif-sur-Yvette, Cedex, France;
*Present address: Institut für Zoologie, Technische Hochschule, Schnittspahnstraße 3, D-6100 Darmstadt, FRG

The structure and development of the telencephalon of actinopterygian fishes differs markedly from that in other vertebrates. During development the dorsal parts of the thickened lateral walls of the embryonic telencephalon undergo a process of eversion (Gage 1893; Nieuwenhuys 1962a, 1963) unlike the situation in other vertebrates, where an inversion followed by an evagination of the lateral walls occurs (Holmgren 1922). The degree of eversion and the extent of telencephalic differentiation differs greatly between and within the actinopterygian superorders (Nieuwenhuys 1963; Northcutt and Braford 1980). Indeed, Nieuwenhuys (1963) notes that the structural variety of the telencephalon within the actinopterygians fishes, which account for over 50% of all vertebrates, is more extensive than within the Chondrichthytes, Dipnoi, Amphibia and Reptilia. A general trend towards increased telencephalic differentiation in actinopterygians can be discerned as one advances from the polypteriform to the teleost fishes. Furthermore, a reduction in olfactory input to the telencephalon apparently accompanies increased telencephalic differentiation (Nieuwenhuys 1963; Northcutt and Braford 1980). This trend is not only apparent between the major actinopterygian superorders, but also within the teleost lineage (Northcutt and Braford 1980). However, the secondary olfactory projections in only a few of the more than 25000 teleost species have been examined with modern tracing techniques and most studies have been carried out on representatives of the euteleost lineage (Table 1). Indeed there have been no recent studies on the olfactory bulb projections in clupeomorphs. Only one elopomorph species, *Gymnothorax funebris* (Scalia and Ebbesson 1971), has been examined, and prior to our recent investigation of the olfactory bulb projections of the weakly electric fish, *Gnathonemus petersii* (Mormyridae), one

Experimental Brain Research Series 19

osteoglossomorph, *Xenomystis nigri* (Nieuwenhuys and Verijdt 1983), had been investigated. The osteoglossomorpha although considered to represent the most primitive group of extant teleosts, possess highly differentiated telencephalons (Weston 1937; Meader 1939; Nieuwenhuys 1962b, 1963).

We undertook our investigation of the central connections of the "classical"[1] olfactory system in *Gnathonemus petersii* to see if a reduction in olfactory input to the telencephalon accompanies the extensive telencephalic differentiation in this species. Only a short summary of the central connections of the "classical" olfactory system in *Gnathonemus* (Fig. 1) will be presented here to facilitate comparisons with the olfactory projections in other actinopterygians, agnathans and chondrichtians. A more detailed account of our findings will be presented elsewhere (Rooney et al., 1989).

Olfactory Bulb Projections in *Gnathonemus petersii*

The secondary olfactory projections in *Gnathonemus* were investigated following a unilateral injection of HRP or cobalt into the olfactory bulb (OB). The olfactory bulbs in *Gnathonemus* are typically pedunculated and connected to the telencephalon by elongated olfactory tracts.

In *Gnathonemus* the olfactory tracts can be separated rostral to their entry point into the telencephalon into medial (MOT) and lateral (LOT) subtracts.

[1] We employ the term "classical" olfactory system after von Bartheld et al. (1984), to differentiate the secondary olfactory bulb connections from those of the nervus terminalis system. During the last five years the central projections of the nervus terminalis have been extensively studied in a number of the lower vertebrate species (for review, see Bartheld and Meyer 1988) and it has been suggested that the nervus terminalis system serves a separate function to that of the "classical" olfactory system (Demski and Northcutt 1983).

Fig. 1 A-G. A-F Drawings of transverse sections through the telencephalon and diencephalon of *Gnathonemus petersii* (levels indicated in sagittal schematic drawing, F) showing the position and extent of secondary olfactory projections. Labeled fibers, *broken lines*; terminal areas, *small dots*; retrogradely labeled neurons, *solid circles*. **G**; simplified schematic drawing of efferent fiber connections of the olfactory bulb in *Gnathonemus petersii*. Abbreviations as in list

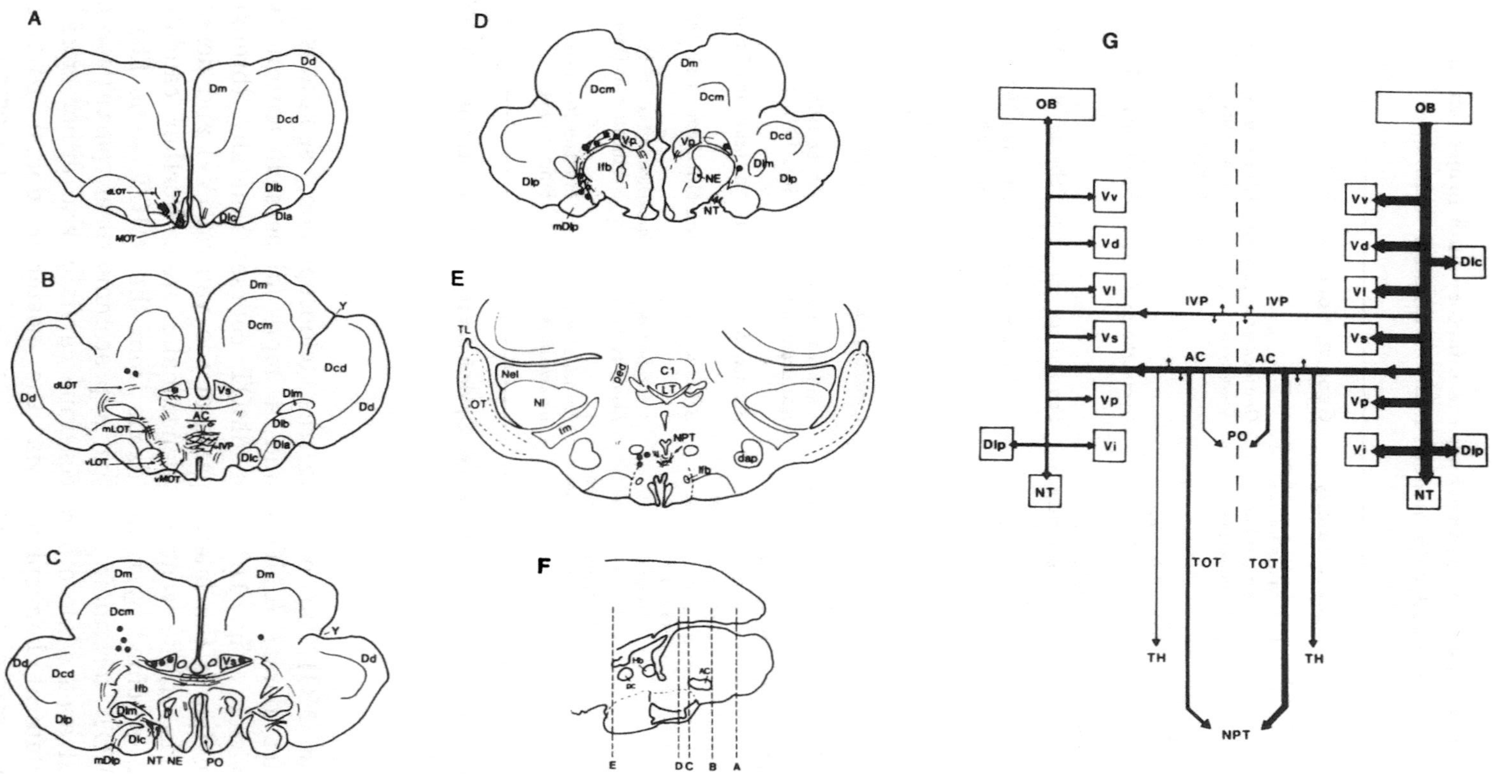
A
Dd
Dm
Dcd
dLOT
Dlb
Dlc
Dla
MOT
B
Dm
Y
Dcm
Dcd
Dd
dLOT
Vs
AC
Dlm
Dd
mLOT
Dlb
IVP
Dla
vLOT
Dlc
vMOT
C
Dm
Dm
Dcm
Y
Dd
Dcd
Vs
Dd
Ifb
Dlp
Dlm
Dlc
mDlp
NT
NE
PO
D
Dm
Dcm
Dcm
Vp
Vp
Dcd
Ifb
NE
Dlm
Dlp
Dlp
NT
mDlp
E
TL
C1
Nel
ped
LT
OT
Nl
Im
NPT
dap
Ifb
F
Hb
DC
AC
E
D
C
B
A
G
OB
OB
Vv
Vv
Vd
Vd
Dlc
Vl
Vl
IVP
IVP
Vs
Vs
AC
AC
Vp
Vp
PO
Dlp
Vi
Vi
Dlp
NT
NT
TOT
TOT
TH
TH
NPT

Table 1. Teleost species in which the olfactory bulb projections have been studied with modern neuroanatomical techniques.

Lineage	Order, Family	Genera, Species	References	Method
Osteoglossomorpha				
	Osteoglossiformes			
	Mormyridae	*Gnathonemus petersii*	Rooney et al. 1989	C,H
	Notopteridae	*Xenomystis nigri*	Niewenhuys and Verijdt 1983	D,H
Elopomorpha				
	Anguilliformes			
	Muraenidae	*Gymnothorax funebris*	Scalia and Ebbesson 1971	D
Euteleostei (Ostariophysi)	Characiformes			
	Serrasalamidae	*Serrasalmus nattereri*	Ebbesson et al. 1981	D,H
	Cypriniformes			
	Cyprinidae	*Carassius auratus*	Ichikawa 1975	D
			Oka 1980	D,P
			Levine and Dethier 1985	H
			von Bartheld et al. 1984	H
			Demski and Northcutt 1983	H
		Cyprinus carpio	Ito 1973	D
	Siluriformes			
	Ictaluridae	*Ictalurus nebulosus*	Finger 1975	D
		Ictalurus punctatus	Bass 1981a,b	A,H,D
(Salmonidae)	Salmoniformes			
	Salmonidae	*Salmonidae gairdneri*	Northcutt and Davis 1983	D
(Percomorpha)	Scorpaeniformes			
	Scorpaenidae	*Sebasticus marmoratus*	Murakami et al. 1983	H,D
	Perciformes			
	Belontiidae	*Macropodus opercularis*	Davis et al. 1981	A,D
	Centrarchidae	*Lepomis cyanellus*	Northcutt and Davis 1983	D
	Pleuronectiformes			
	Pleuronectidae	*Pseudopleuronectes americanus*	Prasada Rao and Finger 1984	D,H

A, autoradiography; C, cobalt lysine; D, degeneration; H, horseradish peroxidase; p, primuline

The MOT, on entering the telencephalon, separates into a compact dorsal (dMOT) and a more dispersed ventral (vMOT) subdivision (Fig. 1A). The dMOT courses caudally through and terminates in the lateral parts of the ventral (Vv) and dorsal (Vd) subfields of the ventral telencephalic area. Further caudally, the dMOT is joined by rising axons from the vMOT, some fibers from the MOT terminate in the supracommissural part (Vs) of the ventral area while others continue caudally and either cross to the contralateral hemisphere via the posterior dorsal part of the anterior commissure (AC) or terminate in the intermediate (Vi) and

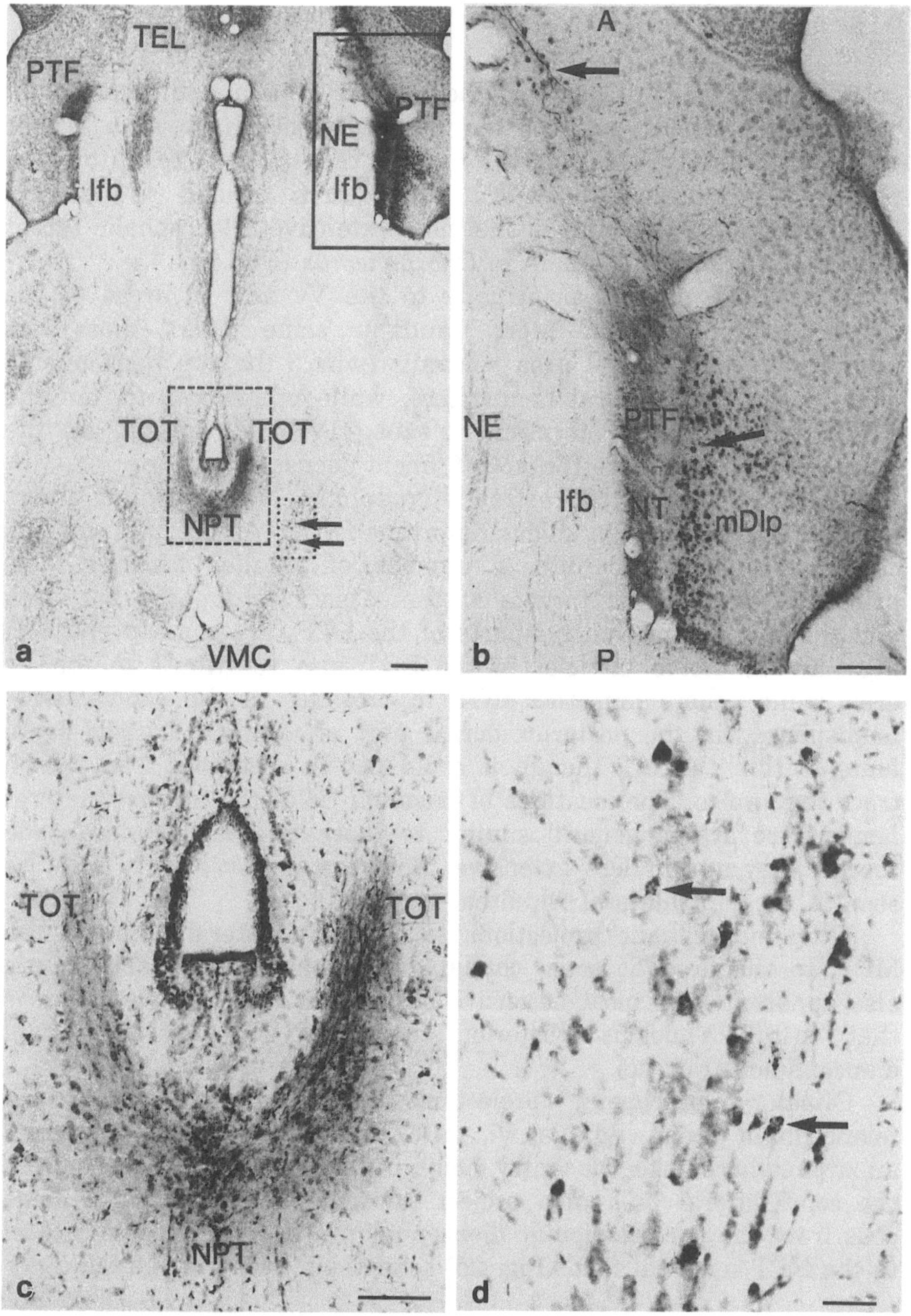

Fig. 2 A-D. Labeling in the telencephalon and diencephalon of the brain of *Gnathonemus petersii* following unilateral HRP injection into the right olfactory bulb. **A** Horizontal section shows position of posterior telencephalic terminal area (PTF) arching the lateral forebrain bundle. Note the weaker contralateral labeling. **Bar** = 200 µm. **B** Detail of Fig. 2a (indicated by *solid rectangle*) at higher magnification showing labeled neurons in the telencephalon. **Bar** = 100 µm. **C** Detail of Fig. 2a (indicated by *dashed rectangle*) at higher magnification showing dense terminal labeling in the nucleus posterior tuberis (NPT). **Bar** = 100 µm. **D** High power photomicrograph of retrogradely filled cells located laterally (*dotted rectangle* in Fig. 2a) to the NPT terminal field in the diencephalon. **Bar** = 25 µm. Abbreviations as in list

postcommissural (Vp) parts of the ventral area and in the medial part of the lateral zone of the dorsal telencephalic area. These post-AC terminal areas plus the nucleus taeniae (NT) constitute the posterior terminal field (PTF) which arcs around the lateral forebrain bundle (lfb) and is the most extensive telencephalic target of descending olfactory fibers in *Gnathonemus* (Fig. 2a).

The vMOT projects principally to the Vv and Vl areas of the rostral area ventralis. More caudally some fibers cross the preanterior commissural area ventralis (where the two hemispheres join) to the contralateral hemisphere, while others rise to join the dMOT. An extensive terminal plexus (IVP) is present in the interventral area where the vMOT fibers decussate.

The LOT enters the rostral telencephalon as a sheet of fibers which curve around the sulcus entorhinalis (E). As the LOT courses caudally it separates into a number of smaller fascicles. The principle telencephalic targets of these fascicles are the Dlc area and the lateral and ventral parts of the PTF (areas mDlp and NT; Figs. 1, 2b). Some fascicles of the LOT also contribute to the Vs area while others join the MOT in crossing to the contralateral hemisphere via the posterior dorsal part of the AC. Endings were found within the AC, though it could not be discerned from which tract they arose. The locations of terminal fields in the contralateral hemisphere are in general similar to those on the ipsilateral side though they appear less extensive. However, no projections could be seen in the contralateral Dlc area.

Extratelencephalic projections to the contralateral bulb via the MOT, to the preoptic area, caudal diencephalon and thalamus are also apparent. The most extensive descending olfactory projection is that to the nucleus posterior tuberis (NPT) in the caudal diencephalon (Fig. 2c).

Olfactory bulbopetal projections were identified from six telencephalic areas, Vd, Vs, Vi, mDc, mDlp (Figs. 1, 2b) and from an area adjacent to the entry point of the LOT. A few cells from the contralateral bulb also project towards the OB. Finally, some cells from the periventricular diencephalon and from an area lateral to the NPT terminal field (Fig. 2d) send their axons to the OB.

Comparison of the Olfactory Projections in *Gnathonemus petersii* with Those of Other Actinopterygian Fishes

Historically the telencephalon of actinopterygian fishes was considered to be dominated by olfaction (Ariëns Kappers et al. 1936). However, neuroanatomical studies during the last 20 years have established that only limited areas of the telencephalon receive olfactory input (Table 1). The restricted nature of the secondary olfactory projections to the telencephalon in *Gnathonemus* closely resembles those in *Xenomystis*, another osteoglossomorph (R. Nieuwenhuys, personal communication). This appears to support the suggestion that a reduction in olfactory input to the telencephalon accompanies increased telencephalic differentiation in actinopterygians (Nieuwenhuys 1963; Northcutt and Braford 1980). However, while Northcutt and Braford (1980) note a trend towards increased telencephalic differentiation accompanied by a reduction in the size of the principle areas dorsalis olfactory target (Dp) between *Salmo* and *Lepomis*, a primitive and advanced euteleost species, respectively, neither the degree of telencephalic differentiation (Nieuwenhuys 1963; Northcutt and Braford 1980) nor the extent of reduction of olfactory input to the telencephalon follows the currently accepted systematic scale for teleosts (Lauder and Liem 1983). In addition, recent studies show a number of other notable interspecies differences in the secondary olfactory projections in actinopterygians (Fig. 3B,C) some of which are discussed below.

Although in most teleost species a differentiation of the MOT into distinct subtracts is reported, the degree of differentiation recognized in the LOT varies between species: in *Ictalurus* (Finger 1975; Bass 1981a), *Sebastiscus* (Murakami et al. 1983), *Pseudopleuronectes* (Prasada Rao and Finger 1984) and *Gnathonemus*, the LOT can be separated into two or more subtracts. In all other teleost species studied with modern tracing techniques, the LOT is reported to be undifferentiated. The olfactory tract fibers decussate via the anterior commissure to the contralateral hemisphere in all teleost species. Both the LOT and MOT cross the AC in all species except *Gymnothorax funebris* (Scalia and Ebbesson 1971) where only the MOT is reported to cross. In *Carassius* (von Bartheld et al. 1984; Levine and Dethier 1985) olfactory fibers decussate via the rostroventral and caudodorsal divisions of the AC. The rostroventral fiber system in *Carassius* may correspond to the pre-anterior commissural

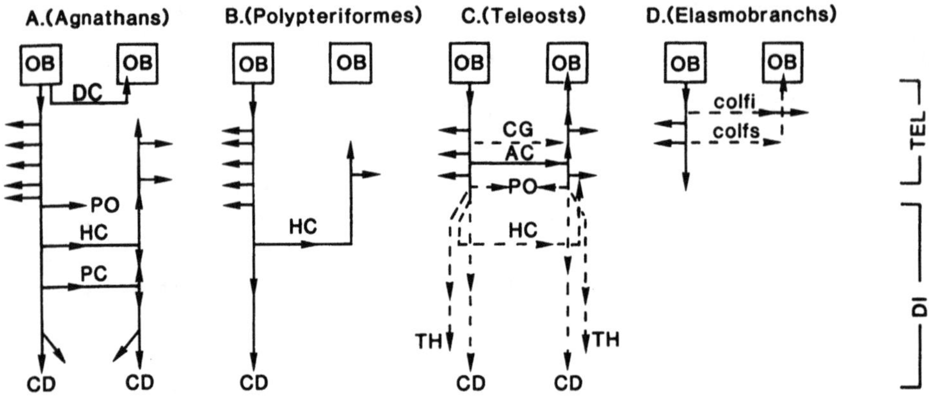

Fig. 3 A-D. Schematic drawings of olfactory bulb projections in **A** Agnathans *(Ichthyomyzon)*, **B** Polypteriformes *(Polypterus)*, **C** Teleosts and **D** Elasmobranchs *(Scyliorhinus)*. The *broken lines* in **C** and **D** indicate projections present only in some of the species. Abbreviations as in list

interventral pathway found in *Gnathonemus* and to the interbulbar commissure of Goldstein described in *Cyprinus* (Ito 1973), *Ictalurus* (Finger 1975; Bass 1981a), *Macropodus* (Davis et al. 1981) and *Pseudopleuronectes* (Prasada Rao and Finger 1984). Only one decussating fiber system in the telencephalon, via the AC, has been reported in other teleost species. Another olfactory pathway to the contralateral hemisphere via the habenular commissure has been demonstrated in *Ictalurus* (Finger 1975; Bass 1981a), *Carassius* (von Bartheld et al. 1984; Levine and Dethier 1985), *Macropodus* (Davis et al. 1981), *Salmo* and *Lepomis* (Northcutt and Davis 1983) and in *Gadus morhua* (own unpublished observation) and is present in several Polypteriform, Chondrostean and Holostean species (see below). This habenular decussating fiber system has not been observed in *Gymnothorax* (Scalia and Ebbesson 1971), *Serrasalmus* (Ebbesson et al. 1981), *Sebastiscus* (Murakami et al. 1983), *Pseudopleuronectes* (Prasada Rao and Finger 1984) or in *Gnathonemus*. The absence of olfactory projections crossing the habenular commissure in these species represents a significant interspecies difference in the secondary olfactory projections in actinopterygian fish. The presence of direct bilateral bulbar projections to the region of the nucleus posterior tuberis in the caudal diencephalon in *Gnathonemus* conforms with most previous studies. Such a projection however is apparently absent in *Gymnothorax funebris* (Scalia and Ebbesson 1971), the only elopomorph species in which the secondary olfactory projections have been studied to date.

In *Gnathonemus* as in *Carassius* (von Bartheld et al. 1984), *Pseudopleuronectes* and *Ictalurus* (Prasada Rao and Finger 1984) bulbopetal neurons were found bilaterally in the telencephalon. Earlier studies (Oka 1980; Ebbesson et al. 1981; Bass 1981b; Murakami et al. 1983) identified such neurons only in the ipsilateral telencephalon. In most recent studies (Bass 1981b; Prasada Rao and Finger 1984; von Bartheld et al. 1984) neurons afferent to the olfactory bulb in the contralateral bulb were identified. Bulbopetal cells in lower brain centers have been reported previously in *Pseudopleuronectes* (Prasada Rao and Finger 1984) and *Carassius* (Levine and Dethier 1985). The cell group in the area lateral to the NPT terminal in *Gnathonemus* (Fig. 2d) may be comparable to the cell group identified in the lateral nucleus preglomerulosus of goldfish by Levine and Dethier (1985). In both *Pseudopleuronectes* (Prasada Rao and Finger 1984) and *Carassius* (Levine and Dethier 1985), the locus coeruleus and raphe nuclei project to the bulb. Such projections were not apparent in *Gnathonemus* or in previous studies in other teleost species.

The Polypteriformes are the most primitive extant actinopterygian fishes and are currently classified as being a specialized order of the Chondrostei (Lauder and Liem 1983). The central connections of the olfactory bulb in the bichir *(Polypterus palmas)* have recently been reexamined with HRP and cobalt tracing techniques (von Bartheld and Meyer 1986; see Fig. 3B) since the results of earlier studies (Nieuwenhuys 1963; Braford and Northcutt 1974; Northcutt and Braford 1980) appeared to be controversial. In this fish the olfactory bulb projects via the medial and lateral olfactory tracts to restricted areas of the ipsilateral pallium and subpallium and to the contralateral subpallium. The MOT appears to furnish most of the fibers which decussate via the habenular commissure to the contralateral hemisphere - unlike the situation in teleosts where the LOT provides the major tract which crosses the habenular commissure. No olfactory projections crossing the anterior commissure or projections to the contralateral bulb have been reported in *Polypterus*. Only an ipsilateral projection to the caudal diencephalon has been shown (von Bartheld and Meyer 1986) in contrast to the situation in teleosts where this projection is reported to descend bilaterally. Neurons afferent to the bulb were identified in the ipsilateral telencephalon and the preoptic area, although projections from more caudal areas or from the contralateral hemisphere could not be detected.

The olfactory projections in the shovel-nosed sturgeon *Scaphirhynchus platorhynchus*, a more representative Chondrostean, and in the long-nosed gar *Lepisosteus osseus*, a Holostean, have been studied with silver degeneration techniques (Northcutt and Braford 1980; Northcutt and Davis 1983). In both species the general pattern of secondary olfactory projections resembles the basic teleost plan. The olfactory bulb projects bilaterally to localized areas of the dorsal and ventral telencephalon, with ipsilateral projections being more extensive than those on the contralateral side. Olfactory fibers cross to the contralateral hemisphere via both the anterior and habenular commissures as in most teleosts. A bilateral olfactory projection to the caudal diencephalon and a projection to the contralateral bulb were also reported in both species. However, as mentioned in the introductory paragraph, a trend towards a reduction in the size of the principle area dorsalis olfactory terminal area in the telencephalon (Dp) was noted between *Scaphirhynchus* and *Lepisosteus* and between *Lepisosteus* and the teleosts *Salmo* and *Lepomis*. At the present time no information is available on the projections from the brain towards the bulb in either *Scaphirhynchus* or *Lepisosteus*.

Olfactory Projections in Agnathans

The secondary olfactory projections in two agnathan species, the Pacific hagfish *(Eptatretus stouti)* and silver lamprey *(Ichthyomyzon unicuspis)* have recently been investigated by Northcutt (1985) and Northcutt and Puzdrowski (1988), respectively. Olfactory input to the telencephalon in these species appears to be more extensive than that observed in gnathostomes. Contralateral projections via the dorsal, habenular and posterior commissures in *Ichthyomyzon* (Fig. 3A) and via the anterior and habenular commissures in *Eptatretus* were identified. Both species show a direct bilateral projection to the caudal diencephalon as has been observed in most actinopterygians. An interbulbar connection via the dorsal commissure is reported in *Ichthyomyzon* though not mentioned in the preliminary report on olfactory projections in *Eptatretus* (Northcutt 1985). Neurons afferent to the olfactory bulb in *Ichthyomyzon* are reported in the ipsilateral telencephalon, preoptic areas and midbrain, as well as in the contralateral bulb. Bilateral telencephalic projections to the bulb as reported in some teleost species and mammals (Broadwell and Jacobwitz 1976) are absent in *Ichthyomyzon*.

Olfactory Projections in Chondrichtyans

The secondary olfactory projections in elasmobranchs (Fig. 3D) appear to be much more restricted than in the agnathans or actinopterygian fishes. However, only a small sample of elasmobranch species have been investigated to date and no information is currently available on the olfactory projections in the holocephalons (chimeras), which together with the elasmobranchs make up the class Chondrichthyes. Studies by Ebbesson and Heimer (1970) on the olfactory projections in the nurse shark *(Ginglymostoma cirratum)* identified only restricted ipsilateral olfactory projections to the ventrolateral telencephalon; no evidence of a contralateral projection or direct projections from the bulb to the caudal diencephalon were apparent. Later studies by Bodznick and Northcutt (1979) on the horn shark (*Heterodontus francisci*) and by Smeets (1983) on spotted dogfish *(Scyliorhinus canicola)* and thornback rays *(Raja clavata)* confirmed that only limited areas of the telencephalon in elasmobranchs receive direct olfactory input. Smeets, (1983), however noted that the secondary olfactory projections in *Scyliorhinus* (Fig. 3D) were more extensive than previously reported in other elasmobranch species, with direct projections to the contralateral telencephalon and contralateral bulb and with distinct subdivisions of the olfactory tracts into lateral and medial components. All studies to date on the olfactory projections in elasmobranchs have employed degeneration techniques, and further studies employing other more recent tracing methods are needed to ascertain other significant interspecies differences and the existence of neurons projecting to the olfactory bulb.

General Discussion

Space does not permit a detailed survey of the olfactory projections in tetrapods and allows for only a few general comments on the complex questions concerning possible homologies between the subdivisions of the everted actinopterygian telencephalon with those of other vertebrates and on the equally vexing problem as to which olfactory projections are plesiomorphic and which are apomorphic.

Considerable controversy surrounds the question whether the search for homologies between the telencephalic subdivisions in actinopterygians and those of other vertebrates is justified (Northcutt and Braford 1980; Northcutt and Davis 1983). Most disagreement between proponents of the search for homologies concerns the identification of which area dorsalis subdivision of the actinopterygian telencephalon corresponds to the medial pallium (MP) or medial cortex (MC) of other vertebrates (Northcutt and Braford 1980; Murakami et al. 1983). Both research groups support the homology between the main posterior area dorsalis olfactory target in actinopterygians (Dp, mDlp in *Gnathonemus*) with the lateral pallium (LP) of amphibians or lateral cortex (LC) of reptiles, respectively. On the basis of topographical position, histochemistry and the eversion hypothesis, Northcutt and Braford (1980) suggest that the ventral and posterior parts of the Dl telencephalic area is the most likely homologue to the MP/MC of other vertebrates. Murakami et al. (1983) disagree and suggest that the ventral Dm area and not the Dlv/Dlp areas correspond to the MP/MC; their proposal is based on the higher order olfactory projections from the LP/LC to the MP/MC in amphibians (Scalia 1976) and reptiles (Halpern 1980). In *Sebastiscus* they found that Dp (LP) projects to ventral Dm and not Dlv/Dlp as would be expected if Dlv/Dlp was the homologue to the MP/MC as proposed by Northcutt and Braford (1980). As the study by Murakami et al. (1983) is the only report on the higher order olfactory projections in a teleost species, further studies on these and other telencephalic connections in other actinopterygian species are required before any definitive conclusions can be drawn on possible homologies of actinoperygian telencephalic subdivisions with forebrain areas in other vertebrates.

With respect to the problem as to which olfactory projections are plesiomorphic and which are apomorphic this brief general survey of the secondary olfactory projections in but a few lower vertebrate species demonstrates the complexity of the puzzle (Fig. 3). To take but one example, direct olfactory bulb projections to the caudal diencephalon appear to be absent in tetrapods as in cartilaginous fishes. Northcutt and Puzdrowski (1988) suggest that direct bulbar projections to the caudal diencephalon may be plesiomorphic for vertebrates and that they have been independently lost in elasmobranchs and tetrapods. If one accepts this suggestion, the apparent absence of a projection to the caudal diencephalon in a teleost *(Gymnothorax funebris*; Scalia and Ebbesson 1971) and the presence of only an ipsilateral projection in *Polypterus* (von Bartheld and Meyer 1986) takes on increased evolutionary significance. Neuroanatomical studies on the olfactory projections in other elopomorph species with modern tracing methods seems necessary to identify whether the apparent absence of the diencephalic projection in *Gymnothorax* is representative of the elopomorph lineage as a whole.

Finally, recent studies on agnathans would support the suggestion that decreased olfactory dominance of the telencephalon accompanied increased telencephalic differentiation during evolution (Nieuwenhuys 1963, Northcutt and Braford 1980). Our results in *Gnathonemus* would support such a trend in actinopterygian fishes.

Acknowledgements. This investigation was supported by grants from the Royal Society European Exchange Fellowship Programme and the Foundation pour la Recherche Medicale Française to D.J.R.

Abbreviations

AC, anterior commissure; **CD**, caudal diencephalon; **ĊG**, commissure of Goldstein; **Cl**, lobulus corporis cerebelli; **colfi**, commissura olfactoria inferior; **colfs**, commissura olfactoria superior; **CP**, posterior commissure; **D**, area dorsalis telencephali; **dap**, nucleus dorsalis anterior pretectalis; **DC**, dorsal commissure; **Dcd**, dorsal central zone of area dorsalis; **Dcm**, medial central zone of area dorsalis; **Dd**, dorsal zone of area dorsalis; **Dl**, lateral zone of area dorsalis; **Dla,b,c**, subdivisions of the lateral zone of area dorsalis (Weston 1937); **Dlm**, medial part of the lateral zone of area dorsalis; **dLOT**, dorsal fascicles of lateral olfactory tract; **Dlp**, posterior part of the lateral zone of area dorsalis; **Dm**, medial zone of area dorsalis; **Dp**, posterior zone of area dorsalis; **lfb**, lateral forebrain bundle; **Hb**, nucleus habenulae; **HC**, habenular commissure; **IT**, intermediate tract; **IVP**, interventral plexus; **LT**, lobus transitorius cerebelli; **mDlp**, medial part of lateral zone of area dorsalis; **mLOT**, medial fascicles of lateral olfactory tract; **MOT**, medial olfactory tract; **NE**, entopeduncular nucleus; **Nel**, nucleus exterolateralis of torus semicircularis; **Nl**, nucleus lateralis of torus semicircularis; **NPT**, nucleus posterior tuberis; **NT**, nucleus taeniae; **OB**, olfactory bulb; **OT**, optic tectum; **PC**, posterior commissure; **ped**, cerebellar peduncle; **PO**, preoptic area; **PTF**, posterior terminal field; **TEL**, telencephalon; **TH**, thalamus; **TL**, torus longitudinalis; **tm**, Marklager of tectum; **TOT**, tubero olfactory tract; **V**, area ventralis telencephali; **Vd**, dorsal part of area ventralis; **Vi**, intermediate part of area ventralis; **Vl**, lateral part of area ventralis; **vLOT**, ventral fascicles of lateral olfactory tract; **Vm**, medial part of area ventralis; **VMC**, ventral mesencephalic commissure; **vMOT**, ventral fascicles of medial olfactory tract; **Vp**, postcommissural part of area ventralis; **Vs**, supracommissural part of area ventralis; **Vv**, ventral part of area ventralis; **Y**, sulcus ypsiliformis

References

Ariëns Kappers CU, Huber GC, Crosby EC (1936) The comparative anatomy of the nervous system of vertebrates, including man. MacMillan, New York

Bass AH (1981a) Olfactory bulb efferents in the channel catfish *Ictalurus punctatus*. J Morphol 169:91-111

Bass AH (1981b) Telencephalic efferents in the channel catfish *Ictalurus punctatus*: projections to the olfactory bulb and optic tectum. Brain Behav Evol 19:1-16

Bodznick D, Northcutt RG (1979) Some connections of the lateral olfactory area of the horn shark *(Heterodontus francisci)*. Soc Neurosci; Abstr 5:139

Braford MR, Northcutt RG (1974) Olfactory bulb projections in the bichir, *Polypterus*. J Comp Neurol 156:165-178

Broadwell RD, Jacobwitz DM (1976) Olfactory relationships of the telencephalon and diencephalon in the rabbit. J Comp Neurol 170:321-346

Davis RE, Chase R, Morris J, Kaufman B (1981) Telencephalon of the teleost *Macropodus*: experimental localization of secondary olfactory areas and of components of the lateral forebrain bundle. Behav Neurol Biol 33:125-142

Demski LS, Northcutt RG (1983) The terminal nerve: a new chemosensory system in vertebrates? Science 220:435-437

Finger TE (1975) The distribution of the olfactory tract in the bullhead catfish, *Ictalurus nebulosus*. J Comp Neurol 161:125-142

Gage SP (1983) The brain of *Diemyctilus viridescens* from larval to adult life and comparison with the brain of *Amia* and of *Petromyzon*. Wilder Quarter Century Book, Ithaka, pp 259-314

Halpern M (1980) The telencephalon of snakes. In: Ebbesson SOE (ed) Comparative neurology of the telencephalon. Plenum, New York, pp 257-295

Holmgren N (1922) Points of view concerning forebrain morphology in lower vertebrates. J Comp Neurol 34:391-459

Ichikawa M (1975) The central projections of the olfactory tract in the goldfish, *Carassius auratus*. J Fac Sci Tokyo Univ 13:257-262

Ito H (1973) Normal and experimental studies on the synaptic patterns in the carp telencephalon, with special reference to the secondary olfactory termination. J Hirnforsch 14:237-253

Lauder GV, Liem KF (1983) The evolution and interrelationship of the actinopterygian fishes. Bull Mus Comp Zool 150:95-197

Levine RL, Dethier S (1985) The connections between the olfactory bulb and the brain in the goldfish. J Comp Neurol 237:427-447

Murakami T, Morita Y, Ito H (1983) Extrinsic and intrinsic fiber connections of the telencephalon in a teleost, *Sebastiscus marmoratus*. J Comp Neurol 216:115-131

Meader RE (1939) The forebrain of bony fishes. Proc Kon Ned Akad Wet Sci 42:657-670

Nieuwenhuys R (1962a) The morphogenesis and the general structure of the actinopterygian forebrain. Acta Morphol Neerl Scand 5:65-78

Nieuwenhuys R (1962b) Trends in the evolution of the actinopterygian forebrain. J Morphol 3:69-88

Nieuwenhuys R (1963) The comparative anatomy of the actinopterygian forebrain. J Hirnforsch 6:171-192

Nieuwenhuys R, Verijdt PWY (1983) The telencephalon of the teleost fish *Xenomystis nigri*. Acta Morphol Neerl Scand 21, 244

Northcutt RG (1985) Efferent projections of the olfactory bulbs in the Pacific hagfish. Am Zool 25:106A

Northcutt RG, Braford MR (1980) New observations on the organization and evolution of the telencephalon in ray finned fishes. In: Ebbesson SOE (ed) Comparative neurology of the telencephalon. Plenum, New York, pp 203-255

Northcutt RG, Davis RE (1983) Telencephalic organization in ray finned fishes. In: Northcutt RG, Davis RE (eds) Fish neurobiology, vol 2. Higher brain areas and functions. University of Michigan Press, Ann Arbor, pp 203-236

Northcutt RG, Puzdrowski RL (1988) Projections of the olfactory bulb and nervus terminalis in the silver lamprey. Brain Behav Evol 32:96-107

Oka Y (1980) The origin of the centrifugal fibers to the olfactory bulb in goldfish. Brain Res 185:215-225

Prasada Rao PF, Finger TE (1984) Asymmetry of the olfactory system in the brain of the winter flounder, *Pseudopleuronectes americanus*. J Comp Neurol 225:492-510

Rooney DJ, New JG, Szabo T, Ravaille-Veron M (1989) Central connections of the olfactory bulb in the weakly electric fish, *Gnathonemus petersii* Cell Tissue Res 257:423-436

Scalia F (1976) Structure of the olfactory and accessory olfactory systems. In: Llinas R, Precht W (eds) Frog neurobiology. Springer, Berlin Heidelberg New York, pp 213-233

Scalia F, Ebbesson SOE (1971) The central projections of the olfactory bulb in a teleost *(Gymnothorax funebris)*. Brain Behav Evol 4:376-399

Scalia F (1976) Structure of the olfactory and accessory olfactory systems. In: Llinas R, Precht W (eds) Frog neurobiology. Springer, Berlin Heidelberg New York, pp 213-233

Scalia F, Ebbesson SOE (1971) The central projections of the olfactory bulb in a teleost *(Gymnothorax funebris)*. Brain Behav Evol 4:376-399

Smeets WJAJ (1983) The secondary olfactory connections in two chondrichthyans, the shark *Scyliorhinus canicula* and the ray, *Raja clavata*. J Comp Neurol 218:334-344

von Bartheld CS, Meyer DL (1986) Central connections of the olfactory bulb in the bichir, *Polypterus palmas*, reexamined. Cell Tissue Res 244:527-535

von Bartheld CS, Meyer DL (1988) Central projections of the nervus terminalis in lampreys, lungfishes, and bichirs. Brain Behav Evol 32:151-159

von Bartheld CS, Meyer DL, Fiebig E, Ebbesson SOE (1984) Central connections of the olfactory bulb in goldfish, *Carassius auratus*. Cell Tissue Res 238:475-487

Weston JK (1937) Notes on the telencephalon of *Mormyrus* and *Gnathonemus*. Proc Kon Ned Akad Wet Sci 40:894-900

Glycoproteins of the Extracellular Matrix Mediate Plasticity in the CNS

Rupert Schmidt[1], Stefan Rother[1]
and Walter K. Schwerdtfeger[2]
[1]Zoologisches Institut der J.W. Goethe-Universität,
Siesmayerstraße 70, 6000 Frankfurt/M., FRG.
[2]Max-Planck-Institut für Hirnforschung, D-6000 Frankfurt/M., FRG.

The anatomical specificity of synaptic connections within neuronal microcircuits of the brain is assumed to code for the contents of acquired information (the semantic aspect of an "engram"). Biochemical reactions, on the other hand, provide the mechanisms that modify the efficacy of individual synapses (the syntax of learning and remembering). Many synapses become temporarily activated when an animal is exposed to new stimuli or when it learns a new behavior. However, the physiological mechanisms that select which new associative pathway will be consolidated and which one will be demolished during long-term memory formation are largely unknown. Teleologically, some additional "factor" might modify all activated synapses if a new behavior proves advantageous. Such a factor should be readily available throughout the central nervous system (CNS). Like, e.g., cell adhesion molecules, it should have the potential to modify fine structural or functional characteristics of synaptic connections, be responsive to the functional state of the neural microenvironment (e.g., to changes in ionic concentrations), and be regulated by a mechanism that evaluates the importance of the preceding activity for the biological fitness of the organism (e.g., hormones or dearousal), in order to determine what has to be consolidated after learning.

We studied the distribution and biochemical characteristics of ependymins that fulfill some of the requirements for a factor involved in regulation of neuronal plasticity. Ependymins are nervous system-specific (Schmidt and Lapp 1987a), acidic (Shashoua 1988; Schmidt 1989), calcium-binding (Schmidt 1986) glycoproteins (37, 32 and 26 kDa apparent molecular weight, MW, on SDS-polyacrylamide gel electrophoresis, SDS-PAGE) sharing immunoreactivity against the HNK-1 antibody and 3-sulfated glucuronic acid in their carbohydrate moieties with cell adhesion molecules of the N-CAM family (Shashoua et al. 1986). The primary

Experimental Brain Research Series 19

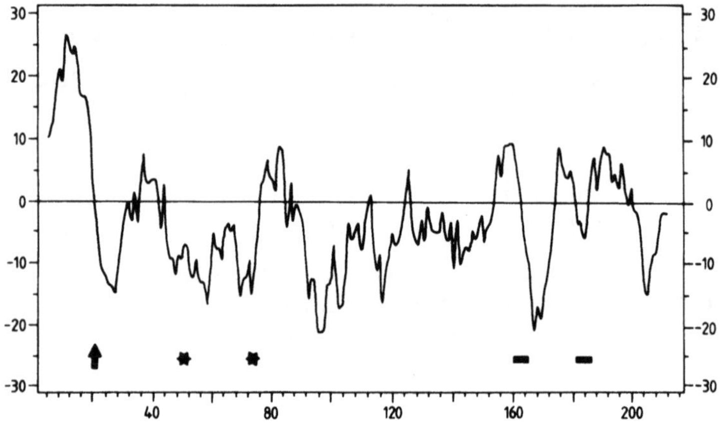

Fig. 1. Hydropathic profile calculated for the amino acid sequence of prepro-ependymin-I (Königstorfer et al. 1989) according to the algorithm SOAP (Kyte and Doolittle 1982; Klein et al. 1985) using an interval of 9 amino acid residues (Gravy = -3.11). The signal peptidase cleavage site is indicated by an *arrow*. To the *left* a characteristic hydrophobic stretch of 21 amino acids will be noted, typical of secretory proteins. *Asterisks* mark two potential N-glycosylation sites; *bars* indicate two clusters of negatively charged amino acids suitable for calcium binding

structure of prepro-ependymin-I, as deduced from cDNA cloning (Königstorfer et al. 1989), comprises a hydrophobic leading sequence of 21 amino acids, typical of secretory proteins (Fig. 1), two N-glycosylation sites, and two clusters of negatively charged amino acids, suitable for calcium binding. Physiologically, the mature proteins occur predominantly as dimers (Schmidt and Shashoua 1981) responding to low calcium concentrations by polymerization (Shashoua 1988). Copurification of an EDTA-sensitive metallo-protease activity with ependymins (Shashoua and Holmquist 1986) suggested that the closely related forms of monomer ependymins (Schmidt and Shashoua 1983) might be derived from each other by proteolysis (but see "Results").

The proteins were discovered by their enhanced synthesis, following vestibulomotoric training forcing goldfish to swim with an attached float (preferential incorporation of radioactive precursor as compared with various controls), purified and used to raise antisera (Shashoua 1976).

They were named ependymins because, immunohistologically, they were first encountered in the periventricular gray of the ependymal zone in cells tentatively identified as glial (Benowitz and Shashoua 1977). In fish, the ependymal zone is rich in clusters of neurosecretory cells intermingled with glial elements and must not be confused with the monocellular ependymal layer typical of mammals. In cultures derived from goldfish optic tectum, ependymins were demonstrated in cells of glial morphology (Majocha et al. 1982), staining neither with antibodies against the nerve growth factor (NGF) nor against the glial fibrillary acidic protein (GFAP). Recently, J.T. Schmidt and Shashoua (1988) described ependymin immunoreactivity in radial glia and ependymal glia cells. On the other hand, ependymins have been demonstrated in cerebellar neurons in goldfish (Schmidt 1989), in pyramidal neurons of the embryonic rat hippocampus in culture (Schmidt et al. 1986), and in small amounts in synaptosomal preparations (Schmidt and Lapp 1987b). Here, we describe distribution of ependymins after light-microscopic and electron-microscopic examination.

Methods

Antisera Ependymins were isolated from goldfish *(Carassius auratus)* brain by chromatography on concanavalin A and SDS-PAGE as described (Schmidt and Shashoua 1981), including protease and esterase inhibitors in the buffer medium (see Shashoua and Holmquist 1986). N-linked carbohydrates were removed with glycopeptidase F (Schmidt 1989). Deglycosylated ependymins were purified on SDS-PAGE. Antisera against glycosylated and deglycosylated ependymins, respectively, were produced in rabbits using standard procedures.

Fixation for Light Microscopy Good fixation was obtained with a modification of Bouin's fixative at 4°C: 1 g of picric acid in 150 ml 80% ethanol and 60 ml 37% formaldehyde diluted with 15 ml acetic acid were mixed immediately before use. Whole juvenile zebra fish *(Brachydanio rerio; Cyprinidae,* kindly provided by Dr. C. Stürmer, Tübingen, FRG) 4 to 35 days posthatching were directly transferred into the fixative. The fixative was diluted out with Tris-buffered saline (TBS: 90 mmol NaCl in 50 mmol tris(hydroxymethyl)aminomethane/HCl, pH 7.4)2 weeks later. The fishes' heads were immersed in 10% and 20% sucrose in TBS until they sunk, mounted on cryostat chucks and cut to 3-6 µm coronal sections at -21°C. Sections were transferred onto microscopic slides (coated with a mixture of 0.5% gelatin, 0.05% $CrK(SO_4)_2 \times 12\ H_2O$ and 0.02% NaN_3 in H_2O) and allowed to stretch for 3 h at 40°C.

Immunofluorescence Staining Anti-ependymin antiserum (raised against deglycosylated ependymin; dilution of 1 : 100) was applied at 4°C for 20 h in TBS containing 3% neutral goat serum. After washing with TBS, fluorescein isothiocyanate conjugated (FITC) goat anti-rabbit IgG was used as second antibody (1 : 40 in TBS, 1% neutral goat serum, 4°C, 20 h), followed by 3 rinses in TBS. To reduce fading of FITC labeling, slides were covered with glycerol containing 10% of a filtered solution of 100 mg p-phenylenediamide in 10 ml phosphate buffered saline (PBS; adjusted to pH

8.0 with $Na_2CO_3/NaHCO_3$) prior to observation in an Olympus IMT2 microscope equipped with high pressure mercury illumination, the BH2-DMBG dichroic mirror unit, excitation filters 20 BG-W-2 and EY-455-W23 and the barrier filter G 520.

Perfusion Fixation For electron-microscopic localization of ependymins, goldfish were kept anesthetized by superfusion of the gills with 0.01% ethyl m-aminobenzoate methanesulfonate via a tube inserted into the fish's mouth. The pectoral skin and muscles were removed to expose the heart. Carefully avoiding air bubbles, the truncus arteriosus was penetrated via the heart ventricle with an injection cannula (outer diameter 0.22 mm) which was tied in place with fine surgical thread. Immediately before commencing perfusion with 30 ml of 200 m*M* PBS, pH 7.2 containing 2670 units heparin sulfate (Sigma; to remove erythrocytes and prevent plugging of blood vessels), the atrium was cut open to allow the blood to extravasate. Perfusion was then continued with 60 ml of 200 m*M* PBS containing 4% paraformaldehyde, 1% glutardialdehyde and 0.2% picric acid (at 5 ml/min; 4°C). Fish developing heart failure or exhibiting incomplete fixation of the gills were discarded. Brains were immediately dissected and transferred into the same fixative. After immersion for 3 days, they were washed in PBS and incubated overnight in PBS containing 10% and 20% sucrose, respectively.

Immunoperoxidase Staining Sections of 50 μm were cut on an Oxford vibratome and incubated in antiserum raised against the 37 kDa ependymin (dilution of 1 : 100 in PBS, 4°C, 20 h). After washing with PBS, the method of Hsu et al. (1981) was used to visualize the antigen-antibody reaction product by coupling to the avidin-biotin-complex (ABC). Sections were incubated for 1 h with biotin-coupled goat anti-rabbit IgG (Vector Laboratories), washed with 0.1 *M* PBS, and incubated for 1 h with avidin and biotinylated peroxidase. After washing, sections were reacted with 0.1% 3,3'-diaminobenzidine and 0.02% H_2O_2 in PBS. Next, sections were postfixed in 4% OsO_4, dehydrated in ethanol and embedded in Durcupan (Fluka) enclosed between layers of transparent plastic foil. After polymerization of the medium (48 h at 60°C), blocks were trimmed for sectioning on a Reichert ultramicrotome. Ultrathin sections were stained with lead citrate and examined with a Zeiss EM 10 electron microscope.

Results

Relationships Between the 37 and 32 kDa Ependymins

Using protease and esterase inhibitors, ependymins were prepared on SDS-PAGE with MWs of 37 and 32 kDa, respectively. By glycopeptidase F treatment the MW of both ependymins was reduced to 28 kDa (Fig. 2g). In some experiments an additional product was revealed from the 37 kDa form, migrating at 32 kDa, and minor bands at approximately 22 kDa. Similar results were

Table 1. Amino acid compositions of ependymins

	Ependymin ß	Ependymin-I	Ependymin-γ	Ependymin-II
Asx	17 ± 1	17	18 ± 1	18
Thr	12 ± 3	11	12 ± 2	10
Ser	17 ± 5	20	18 ± 4	22
Glx	25 ± 2	23	26 ± 2	21
Pro	12 ± 2	11	12 ± 2	11
Gly	21 ± 3	13	22 ± 4	13
Ala	11 ± 1	9	12 ± 2	10
Val	12 ± 2	13	12 ± 2	13
Cys	2 ± 1	5	2 ± 1	5
Met	1 ± 1	5	1 ± 1	3
Ile	6 ± 1	5	6 ± 1	6
Leu	14 ± 2	12	15 ± 1	15
Tyr	5 ± 1	6	4 ± 2	7
Phe	12 ± 2	14	12 ± 2	13
Lys	12 ± 1	11	12 ± 1	11
His	11 ± 2	14	10 ± 2	11
Arg	6 ±11	5	6 ± 1	5
Trp	n.d.	1	n.d.	1

Amino acid compositions deduced from cDNA cloning of prepro-ependymin-I (Königstorfer et al. 1989) and prepro-ependymin-II are compared with the compositions published by Schmidt and Shashoua (1983), recalculated (± standard deviation) under the assumption of MW 21960 (ß, column 2) and 21757 (γ, column 4), respectively, for the mature ependymin polypeptides (i.e., without the leading sequence). The content of tryptophan was not determined in the earlier experiments (n.d.). Data show good correspondence; the only significant differences were detected with cysteine and methionine residues, both of which are very sensitive to oxidation, and in the number of glycine residues probably due to the presence of this amino acid in the buffer system used for preparation, as already discussed by Schmidt and Shashoua (1983).

obtained with endoglycosidase F (not shown). It is inferred that the 37 and 32 kDa ependymins differ mainly in the amount of N-linked glycosylation, in accordance with similarities in the amino acid compositions (Table 1), peptide mapping after limited proteolysis, and binding characteristics to antibodies raised against either of the immunogens (Schmidt and Shashoua 1983). The identical apparent MWs of the deglycosylated proteins do not necessarily implicate, however, that the polypeptide chains are identical. Microheterogenieties revealed by gas-phase sequencing of the N-terminal amino acids rather point out that both ependymins comprise a family of closely related sequences (Königstorfer et al. 1989), confirming earlier results that demonstrated seven isoelectric ependymin variants (Shashoua 1985; Schmidt 1989). In some of the 32 kDa ependymin variants, Asp and Ile have replaced His and

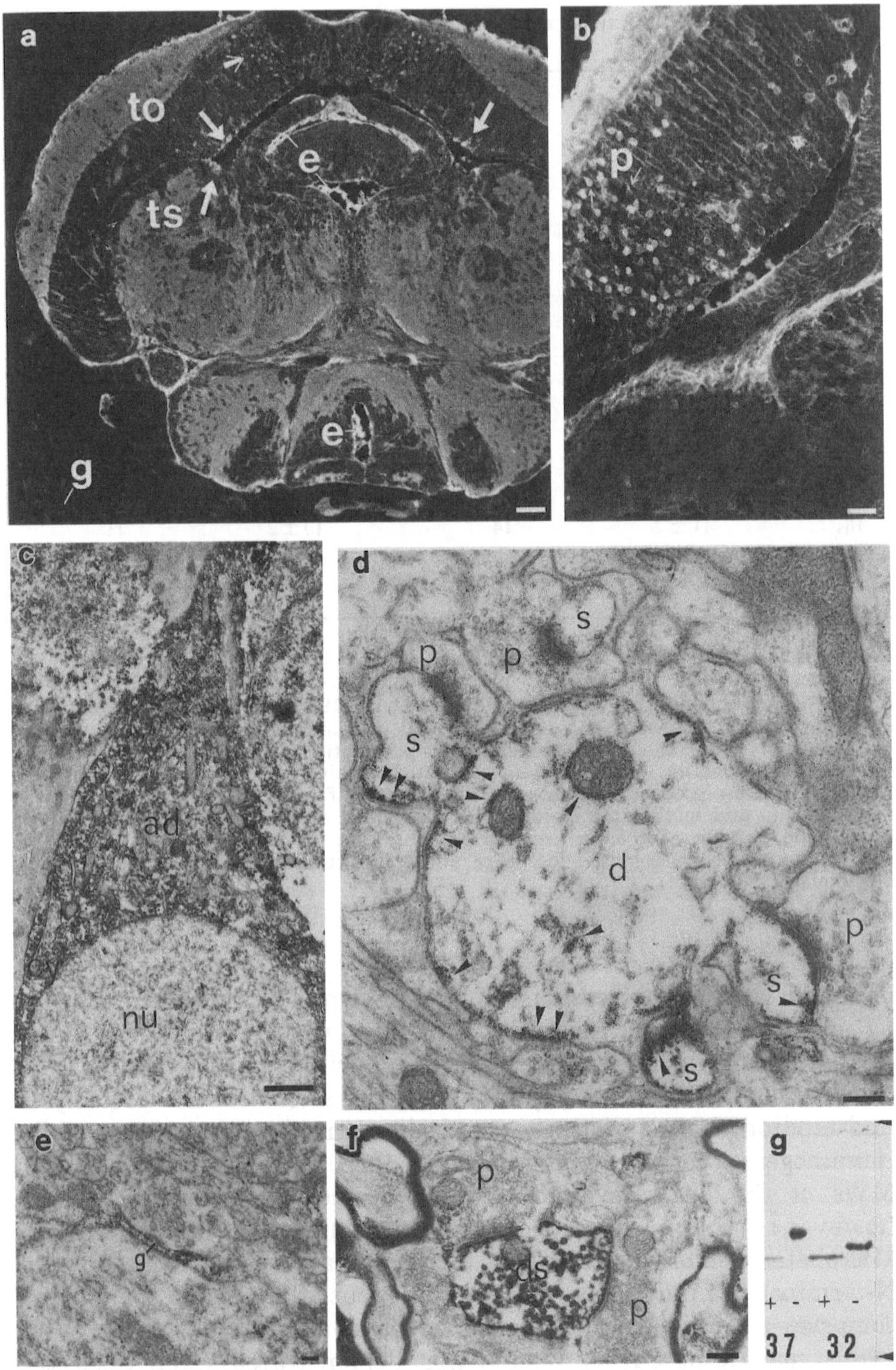
a
to
ts
e
e
g
b
p
c
ad
nu
d
s
p
p
s
d
p
s
s
e
g
f
p
ds
p
g
+ - + -
37 32

Thr, which are the predominant amino acids of the 37 kDa ependymin at position 3 and 13, respectively. These and a few further point mutations between residues 3-66 and 119-153 were also revealed by cDNA cloning of a second precursor, prepro-ependymin-II (A. Königstorfer, S. Sterrer, R. Schmidt and W. Hoffmann, unpublished results). Amino acid compositions deduced from cDNA cloning show good correspondence with earlier analyses of the proteins (Table 1). Calculated MWs were lower as compared to the apparent MWs on SDS-PAGE. The discrepancy might result from abnormal migration on SDS-PAGE as described for several calcium binding proteins. There is, however, unequivocal evidence that ependymin precursors have been cloned, since a fusion protein constructed from lacZ and 50 amino acid residues of the C-terminal sequence was recognized by anti-ependymin antiserum (A. Königstorfer, S. Sterrer, R. Schmidt and W. Hoffmann, unpublished results).

◄

Fig. 2a-g. a Immunofluorescence staining of a coronal section through the head of a 28-day-old zebra fish. Fluorescence is clearly confined to the CNS. Bone, cartilage, skin, gills (*g*) and muscle tissue do not stain for ependymins. *Arrows* indicate immunoreactive cells close to the surface of the optic ventricle within the optic tectum (*to*) and the dorsal part of the torus semicircularis (*ts*). Some cells of this population have migrated to higher layers of the tectum (*small arrowhead*). Stained ependymal cell layers surrounding the cerebellar and central ventricles give rise to strongly positive material within the extracellular fluid (e). In addition, most of the leptomeninx displays specific staining for ependymins. *Bar*, 40 μm. **b** Immunofluorescence of tectal cells at higher magnification. Lamination of the tectum has not yet fully developed. Individual pyramidal neurons (*p*) stain brightly for ependymins and extend long apical dendrites through the granular cell layer to ramify in superficial molecular layers, where immunofluorescence is less distinct. *Bar*, 20 μm. **c** Electron micrograph of a typical pyramidal neuron in the stratum periventriculare of an adult goldfish. Note intensive immunostaining of the apical dendrite (*ad*), limited width of the somatic cytoplasm (*cy*) and absence of immunoreactivity from the nucleus (*nu*). *Bar*, 1 μm. **d** Marked ependymin-staining (*arrowheads*) of a cross-section through a dendrite (*d*) in the goldfish tectum with several spines (*s*). Subsynaptic densities are apposed by presynaptic boutons (*p*) filled with synaptic vesicles. *Bar*, 0.2 μm. **e** As compared with the neuronal profile shown in **c**, immunoreactivity is much fainter in a glial cell element (*g*). *Bar*, 0.2 μm. **f** Strong immunoreactivity in a dendritic shaft (*ds*) that is contacted by at least two presynaptic elements (*p*). Products of the immunoreaction are deposited at numerus microtubules and a mitochondrion. The dendrite is surrounded by several myelinated fibers displaying little or no immunoreactivity. *Bar*, 0.2 μm. **g** 100 μg of the 37 kDa (*trace 1*) and the 32 kDa (*trace 3*) ependymin, respectively, were incubated for 24 h with 2 units of glycopeptidase F (marked, +) and separated by SDS-PAGE. The undigested ependymin molecules were run on the same gel for comparison (*traces 2 and 4*; marked, -). Obviously, a polypeptide of identical MW (approximately 28 kDa) is obtained from either of the ependymins, suggesting that they differ mainly in their carbohydrate moieties. *Arrowheads* indicate the start position and bromophenol blue used to monitor the migrating front

Ependymins in Juvenile Zebra Fish

As ependymins are involved in neural plasticity, it was of interest to investigate whether they are expressed in early stages of the developing teleost brain, when much remodeling takes place in the CNS. In zebra fish, ependymin-like immunoreactivity could clearly be identified 4 days after hatching (not shown). Figure 2a shows immunofluorescence staining of a coronal section through the head of a 28-day-old zebra fish. The label is strictly confined to the nervous tissue, in accordance with radioimmunoassay (RIA) measurements on goldfish (Schmidt and Lapp 1987a). Bright staining was observed in the optic tectum, the hypothalamus, vagal lobes, the leptomeninx, and large neurons of the cerebellum probably corresponding to Purkinje cells of the mammalian brain. Close to the ependymal cell layers strongly positive material was deposited within the extracellular matrix of the ventricles. At higher magnification (Fig. 2b), intensely labeled cells are seen in the periventricular layer of the optic tectum. By morphological criteria these cells appear to be pyramidal neurons extending long apical dendrites to the marginal layers of the tectum (compare Meek 1983). Some of them are situated at the ventricular surface, others have migrated to more superficial regions of the tectum which will in later stages differentiate to comprise the stratum periventriculare and the stratum album centrale. At light microscopic examination, it was difficult to decide, however, whether the staining was cytoplasmic or whether ependymins had been bound to, or precipitated on, cell membranes. This possibility had to be taken into account, since ependymins are known to be secreted (see "Discussion") and were localized by immunohistological methods in the ventricular fluid when zebra fish (Fig. 2a) or goldfish brains were frozen without fixation in melting 2-methylbutane (Schmidt 1989).

Fine Structural Localization

Electron microscopy was performed on adult goldfish big enough to be perfused with fixative through the heart in order to achieve reasonable fine structural preservation. Figure 2c depicts one of many typical ependymin positive pyramidal neurons in the stratum periventriculare with the basis of its apical dendrite. The ABC deposit is clearly intracellular. The neurons possess a large unstained nucleus surrounded by a small intensely labeled cytoplasmic envelop which cannot easily be resolved by light microscopy (compare Fig. 2a, 2b and Schmidt 1989). Marked ependymin staining is also seen in cross-sections through dendrites (Fig. 2d and 2f), some of which possess dendritic spines typical of pyramidal neurons. The precipitate of the immunoreaction is deposited at various diffusion barriers (cell membranes and subcellular organelles), not allowing to decide what the subcellular distribution of the antigen has been in vivo. Some product of the immunoreaction is also displayed within presynaptic profiles, which may, however, be derived by leakage from postsynaptic elements. As compared with neurons (Fig. 2c) immunoreactivity is much fainter in glial cell elements (Fig. 2e) and mainly confined to regions close to the ventricles.

Discussion

Applying immunohistochemical techniques, we localized ependymins in the juvenile zebra fish and adult goldfish brain mainly in neurons, in particular in pyramidal neurons of the stratum periventriculare in the optic tectum. Staining of glial cells was comparatively rare and might be due to uptake of ependymins into lysosomes. Our results are consistent with earlier observations of ependymins in neurons, e.g., in pyramidal cells of the rat hippocampus (Schmidt et al. 1986), but not necessarily contradictory to other reports on ependymin-like immunoreactivity in glial cells. Various reasons may account for differences in the localization:

1. Ependymins may be expressed in particular types of glial cells in addition to pyramidal neurons.
2. When grown in culture, cells may dedifferentiate and express other protein constituents than in situ.
3. Physiologically, ependymins occur as closely related monomers with similar antigenic characteristics (Schmidt and Shashoua 1983), in several dimeric conformations (Schmidt and Shashoua 1981), and they polymerize in the absence of calcium (Shashoua 1988). They display microheterogenieties in their amino acid compositions and differ in the degree of glycosylation (see "Results" and Königstorfer et al. 1989). It is likely that each of the various constitutions and conformations displays a distinct distribution, and that antisera raised against the respective forms of the molecule recognize a different set of ependymin epitopes. Nevertheless, both antisera used in this study stain pyramidal neurons, although one was raised against a glycosylated (37 kDa) and the other against the deglycosylated immunogen.

4. Differential expression and posttranslational modification of ependymins may depend on the ontogenetic age of the animal.

5. Ependymins may be secreted from one cell population and be taken up by another. In accordance with the hydrophobic leading sequence (Fig. 1) and the cleavage site for a signal peptidase (Königstorfer et al. 1989), several lines of evidence prove that ependymins are secreted in cell cultures (Majocha et al. 1982) and in situ: Following the float training (Shashoua 1979) or an active shock avoidance conditioning in a shuttle box (Schmidt 1989), injected radioactive valine was preferentially recovered in ependymin molecules of the extracellular brain fluid (ECF). The steady-state concentrations of ependymins in the ECF also increased 8 h after learning (Schmidt 1987). Furthermore, secretion of ependymin-like immunoreactivity has recently been demonstrated after long-term potentiation of the rat hippocampus in situ (Fazeli et al. 1988). Based on electron-microscopic studies with the domestic cat, Gonzales-Santander (1979) suggested that proteins like ependymins might be released from secretory ependymocytes by apocrine mechanisms. Presently, however, secretion by an exocytotic mechanism cannot be excluded, as high specific ependymin concentrations were observed in a microsomal fraction containing secretory granules amongst various other vesicular structures (Schmidt and Lapp 1987b).

Secretion might either serve as a drainage device to inactivate ependymins or rather be a step towards their interaction with remote structures. Various experiments support the latter hypothesis: Anti-ependymin antibodies prevented memory consolidation when injected into brain ventricles after acquisition of the float-swimming skill (Shashoua and Moore 1978; Schmidt 1987) or the shock avoidance conditioning (Piront and Schmidt 1988). They also interfered with activity-dependent sharpening of the multiunit receptive fields during regeneration of retinotectal projections (Schmidt and Shashoua 1988). As uptake of antibodies into neurons is very unlikely, anti-ependymins are assumed to inactivate secreted antigen prior to its interaction with unidentified targets or to interfere with ependymin molecules bound to cell membranes. The observation that radioactive ependymins injected in vivo are incorporated into particles sedimenting in the synaptosomal fraction (Schmidt and Lapp 1987b) is consistent with such a mechanism. Also recognition of glycosylated ependymins by antibodies directed against sugar epitopes of cell adhesion molecules (Shashoua at al. 1986) suggests that secreted ependymins might interact with cell membranes exposed to the ECF. Furthermore, we have recently shown that ependymins provide a suitable substrate for outgrowing axons of goldfish retinal ganglion cells in vitro (in collaboration with C. Stürmer, Tübingen, and J. Schmidt, Albany, NY). Interestingly, J. Schmidt and Shashoua (1988) gave morphological evidence for increased ependymin synthesis during regeneration of retinotectal projections.

In conclusion, ependymins are contained in neurons and, after secretion, mediate biochemical or structural modifications in the CNS that are essential for plastic changes triggered by memory consolidation and neuronal regeneration.

Acknowledgements. We are grateful to C.Weindel and R.Krauss for excellent technical assistance, to Dr.M.Schmidt and T.Berberich, Frankfurt, for help with calculation of the hydropathic profile, and to Dr.C.Stürmer, Tübingen, for supply with zebra fish. Part of our experiments was supported by grants from the Deutsche Forschungsgemeinschaft (Schm 478/4-2 and 4-3 to R.S.).

References

Benowitz LI, Shashoua VE (1977) Localization of a brain protein metabolically linked with behavioral plasticity in the goldfish. Brain Res 136:227-242

Fazeli MS, Errington ML, Dolphin AC, Bliss TVP (1988) Proteins of the extracellular fluid of dentate gyrus: presence of proteins S-100 and ependymins but absence of NGF in vivo. Eur J Neurosci 1 (Suppl):90

Gonzales-Santander R (1979) Electron-microscopic study of the secretion of the ependymal cells in the domestic cat (ependymin-ß cells). Acta Anat (Baltimore) 103:266-277

Hsu S-M, Raine L, Fanger H (1981) Use of avidin-biotin-peroxidase complex (ABC) in immunoperoxidase techniques: a comparison between ABC and unlabeled antibody (PAP) procedures. J Histochem Cytochem 29:577-580

Klein P, Kanehisha M, Delisi C (1985) The detection and classification of membrane spanning proteins. Biochim Biophys Acta 815:468-476

Königstorfer A, Sterrer S, Eckerskorn C, Lottspeich F, Schmidt R, Hoffmann W (1989) Molecular characterization of an ependymin precursor from goldfish brain. J Neurochem 52:310-312

Kyte J, Doolittle RF (1982) A simple method for displaying the hydropathic character of a protein. J Mol Biol 157:105-132

Majocha RE, Schmidt R, Shashoua VE (1982) Cultures of zona ependyma cells of goldfish brain: an immunological study of the synthesis and release of ependymins. J Neurosci Res 8:331-342

Meek J (1983) Functional anatomy of the tectum mesencephali of the goldfish. Brain Res Rev 6:274-297

Piront M-L, Schmidt R (1988) Inhibition of long-term memory formation by anti-ependymin antisera after active shock-avoidance learning in goldfish. Brain Res 442:53-62

Schmidt JT, Shashoua VE (1988) Antibodies to ependymin block the sharpening of the regenerating retinotectal projection in goldfish. Brain Res 446:269-284

Schmidt R (1986) Biochemical participation of glycoproteins in memory consolidation after two different training paradigms in goldfish. In: Matthies H (ed) Learning and Memory: mechanisms of information storage in the nervous system. Pergamon, Oxford, pp 213-222 (advances in the bioscience Vol. 59)

Schmidt R (1987) Changes in subcellular distribution of ependymins in goldfish brain induced by learning. J Neurochem 48:1870-1878

Schmidt R (1989) Glycoproteins involved in long lasting plasticity in the teleost brain. In: Rahmann H, Lindauer M (eds) Fundamentals of memory formation: neural plasticity and brain function. Prog Zool 37 (In press)

Schmidt R, Lapp H (1987a) Regional distribution of ependymins in goldfish brain measured by radioimmunoassay. Neurochem Intern 10:383-390

Schmidt R, Lapp H (1987b) Subcellular distribution of ependymins in goldfish brain measured by radioimmunoassay. J Neurochem 48:1862-1869

Schmidt R, Shashoua, VE (1981) A radioimmunoassay for ependymins ß and γ: Two goldfish brain proteins involved in behavioral plasticity. J Neurochem 36:1368-1377

Schmidt R, Shashoua VE (1983) Structural and metabolic relationships between goldfish brain glycoproteins participating in functional plasticity of the central nervous system. J Neurochem 40:652-660

Schmidt R, Löffler F, Müller HW, Seifert W (1986) Immunological cross-reactivity of cultured rat hippocampal neurons with goldfish brain proteins synthesized during memory consolidation. Brain Res 386:245-257

Shashoua VE (1976) Brain metabolism and the acquisition of new behaviors. I. Evidence for specific changes in the pattern of protein synthesis. Brain Res 111:347-364

Shashoua VE (1979) Brain metabolism and the acquisition of new behaviors. III. Evidence for secretion of two proteins into the brain extracellular fluid after training. Brain Res 166:349-358

Shashoua VE (1985) The role of brain extracellular proteins in neuroplasticity and learning. In: Activity-Dependent Synaptic Changes. Cell Mol Neurobiol 5:183-207

Shashoua VE (1988) Monomeric and polymeric forms of ependymin: a brain extracellular glycoprotein implicated in memory consolidation processes. Neurochem Res 13:649-655

Shashoua VE, Holmquist B (1986) Extracellular fluid proteins of goldfish brain: evidence for the presence of proteases and esterases. J Neurochem 47:738-743

Shashoua VE, Moore ME (1978) Effect of antisera to ß and γ goldfish brain proteins on the retention of a newly acquired behavior. Brain Res 148:441-449

Shashoua VE, Daniel PF, Moore ME, Jungalwala FB (1986) Demonstration of glucuronic acid on brain glycoproteins which react with HNK-1 antibody. Biochem Biophys Res Commun 138:902-909

Developmental Organization of the Amphibian Pallium

P. Clairambault and J.F. Timmel
Neuroembryology, Univérsité Paris 7
Paris, France

The amphibian telencephalon is generally considered a primitive structure displaying a very simple architecture which is similar to that of Holocephali and Dipnoi (Clairambault and Capanna 1973; Reiner and Northcutt 1987) and which has probably been inherited from a more primitive and unknown ancestor (Northcutt 1981). It has been tempting to search in the mammalian brain for equivalents of the amphibian telencephalic divisions (Ariens Kappers et al. 1936; Herrick 1948). This long and arduous task led to an attempt at a comprehensive interpretation of the vertebrate forebrain, especially of the dorsal part of the telencephalon, more generally named pallium in nonmammalian vertebrates (the so-called "cortex" according to the earlier nomenclature). During the past few years, it has become clear that this approach failed to explain the evolution of the pallial organization through the vertebrates series. One reason was that only rather few morphogenetic studies have been available.

Attempting to bridge the gap between anatomical descriptions and phylogenetical hypotheses concerning the amphibian brain, we have performed developmental studies of the anuran telencephalon (Clairambault 1976) and, recently, of the pallial fiber connections in a urodele, *Pleurodeles*. The two studies yield evidence of strong developmental similarities between the anuran and the urodelan pallium.

As in the anuran telencephalon, the pallium of mature *Pleurodeles* consists of three longitudinal zones: the medial, dorsal and lateral parts. At the rostral end of the hemisphere, they converge onto the olfactory anterior nucleus. In addition, the lateral pallium shows a small area along the rostro-caudal extent of its ventral border: the epistriatum, (Ariens Kappers 1908) which is now interpreted as anterior amygdala (Veenman, this volume) and is morphogenetically derived from the *stratum matricis* of the lateral pallium (Clairambault 1976).

Experimental Brain Research Series 19

The axonal circuitry of *Ambystoma* telencephalon was extensively investigated by Herrick (1948) with the aid of the Golgi method. More recently improved experimental techniques have revealed important pathways, such as the bilateral afferents from the dorsal thalamus in *Triturus* (Northcutt and Kicliter 1980; Clairambault et al. 1986a; Wicht and Himstedt 1988), and helped to recognize that the urodelan pallium is more differentiated than previously assumed.

In *Rana* (Clairambault 1976), the olfactory bulb is already differentiated at the end of embryonic life (stage 23, Shumway 1940), while the developing lateral pallium is represented by a thick *lamina lateralis* and the medial pallium by the very thin *tela prosencephalica*. Morphogenetic events in the lateral pallium precede those of the medial pallium and, in each pallial part, cytological characteristics appear as early as the beginning of premetamorphosis (stage 25). The two main neuronal types (i.e. multipolar neurons in the medial pallium and piriform neurons in the lateral pallium) begin to differentiate during this period and reach their ultimate stage of maturation by the climax stage (Clairambault and Capanna 1970).

The tract tracing study presented in this chapter represents the first approach towards the elucidation of the development of fiber connections in Urodela. This study revealed the precocious appearance of anatomical links between the olfactory bulb and the lateral pallium and transient fibers in the whole pallium.

Preliminary data have been published on such fibers occurring in the olfacto-habenular system (Timmel and Clairambault 1988). Transient fibers have also been found in other vertebrates, e.g. in the visual system (Ebbesson 1984) and the callosal system of birds (O'Leary et al. 1983). Whereas it is fascinating to encounter such fibers in the study of a developmental system, their phylogenetic and ontogenetic significance is still an enigma.

Organization of the Pallial Connections in Pleurodeles

Injections of 0.5 µl 10% horseradish peroxidase (HRP) in distilled water were given into the rostral part of the dorsal pallium in adult *Pleurodeles*. After a survival time of 2-6 days, brains were removed, fixed in Susa's fluid, embedded in paraffin and transversely or sagittally cut to 10 µm thick sections. HRP deposits were visualized with the aid of the avidin-biotin-complex method (Hsu et al. 1981). For further technical details, see the previous study on the habenula of *Triturus* (Clairambault et al. 1986b).

Afferents of the Dorsal Pallium

Rostral telencephalon. The injections lead to the following distribution of labeled perikarya (Fig. 1): In the ipsilateral main olfactory bulb, many mitral cells are seen. The retrograde filling extends beyond the perikarya into the dendritic arborization. The accessory olfactory bulb remains free of labeling. In those cases in which the HRP injection sites are located close to the anterior olfactory nucleus, some labeled neurons are found in the ipsilateral ventral and lateral parts of the latter nucleus.

Pallium. The whole ipsilateral pallium displays many labeled perikarya (medial, dorsal and lateral pallium). These neurons exhibit a branched dendritic tree and pear shaped somata.

Telencephalic ventro-lateral wall. Scarce HRP-labeled neurons are disseminated in the periventricular cell mass of the anterior amygdala in the ipsilateral hemisphere, suggesting the presence of an amygdalo-pallial pathway.

The General Fibrous Network

Medial wall. In the rostral hemisphere, many fibers travel along the medial wall. They probably associate the septum with the pallium, as reported for *Rana* (Halpern 1972). These rostro-medial fibers are located on Herrick's (1948) *tractus post-olfactorius*, interpreted as an efferent pathway of the anterior olfactory nucleus, and, probably, of all other parts of the rostral pallium.

Lateral wall. The whole lateral wall of the hemisphere shows a massive fibrous network throughout its rostro-caudal extent. The dorsal component of this system, which lies on Herrick's (1948) *tractus olfactorius dorso-lateralis*, reaches the caudal pole of the hemisphere. The more substantial, ventral component joins the stria medullaris, crosses through the habenular commissure and turns to the rostro-ventral part of the contralateral hemisphere where no labeled perikarya are seen. Hence, these contralateral fibers may constitute an efferent pallial pathway.

Remarks

Regarding the circuitry of the dorsal pallium in adult *Pleurodeles*, we can formulate four general statements: (1) The basic olfactory

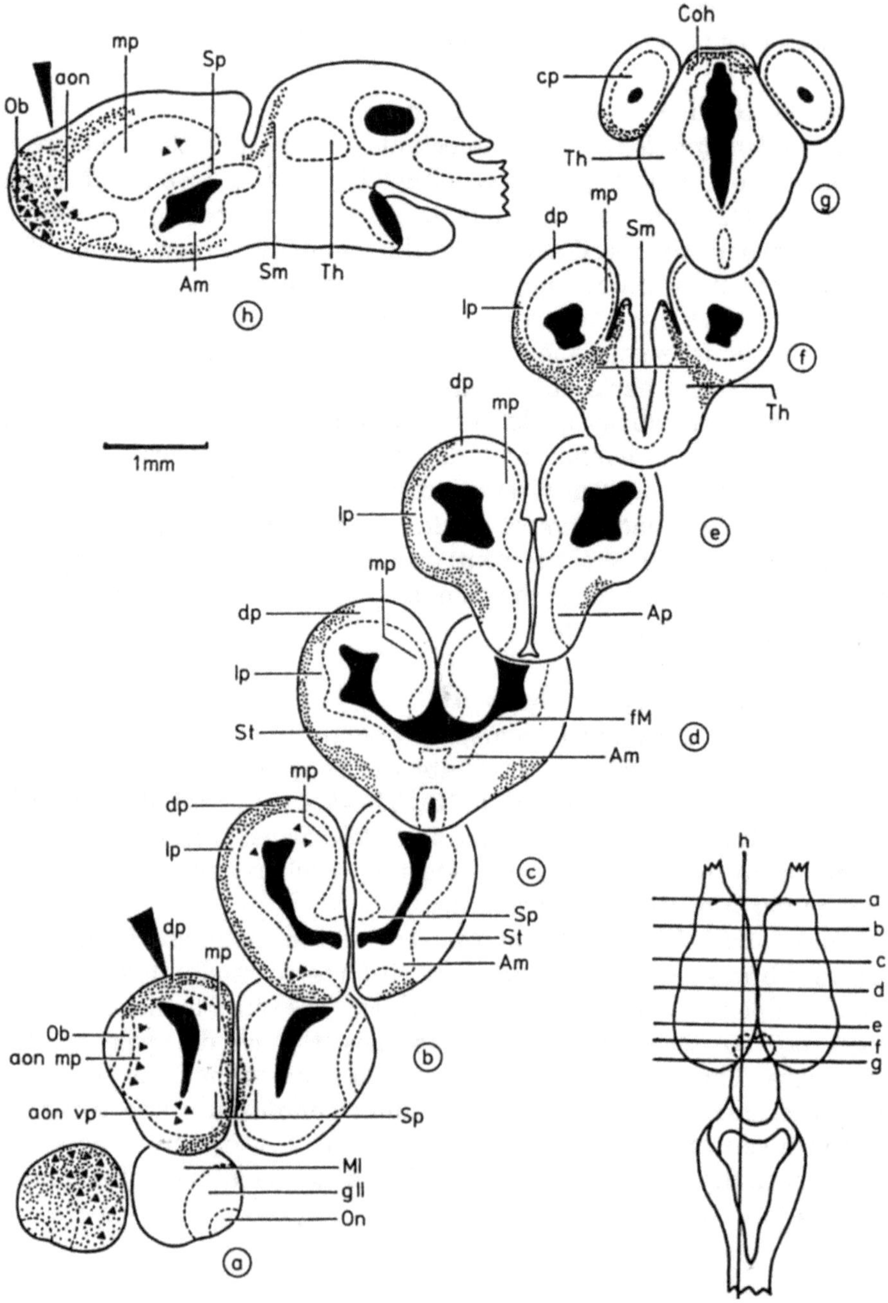

Fig. 1 a-h. Drawings illustrating the connections of the anterior dorsal pallium in adult *Pleurodeles* as revealed by the HRP technique **a-g** transverse sections **h** sagittal section. HRP-filled cells are indicated by *triangles*, fibers by *dotted* areas. *Dotted lines* mark the boundary of the periventricular cell masses. The *large arrowhead* points to the HRP injection site.

pattern described in Anura (Northcutt and Royce 1975) and in *Protopterus* (Reiner and Northcutt 1987) is valid for *Pleurodeles*; (2) The rostral pallial connections represent an intrinsic prosencephalic system; (3) An afferent pathway from the anterior amygdala reaches the rostral pallium. This projection is lacking in *Rana* (Wilczynski and Northcutt 1983); (4) Since no connections were found with the habenula, the latter is probably only related with the caudal pallium, as reported earlier for *Triturus* (Clairambault et al. 1986b).

Development of Pallial Connections in Larval Pleurodeles

Morphogenetic studies yield valuable data which provide a better understanding of the architectural organization of neuronal systems. This work attempts to elucidate the development of pallial fiber connections during the metamorphosis in *Pleurodeles* (Fig. 2). The stages of development of this species were codified by Gallien and Durocher (1957); metamorphosis begins at stage 36 and ends at stage 55. We studied the following stages: 38, 41, 43, 45, 50, 52, 54 and 55. In terms of pallial neuronal connections, metamorphosis can be divided into two parts, the first comprising stage 38 to stage 45, and the second stage 46/47 to stage 55. An external feature marks the temporal limit between the two parts, namely, the barbels (embryonic sensory organs), which disappear at stage 46.

As in the adult, HRP was introduced in the dorsal pallium either by insertion of an HRP crystal or by injection (0.1 µl of a 10% aqueous solution). Due to the small size of the specimens, applications of HRP were always placed in the middle of the dorsal surface of the left hemisphere. Taking this topographic situation into account, can we compare our results in the larvae with those from the adults? Studies of the telencephalic development (Clairambault et al. 1973; Clairambault 1976) give the answer. Two types of evagination are involved in the formation of the amphibian telencephalic hemisphere: a pseudo-evagination of the tela prosencephalica and a true evagination of the prosencephalic lateral wall whose caudo-rostral cell proliferation drives the buildup of the whole hemisphere. This caudo-rostral polarity of the growing hemisphere is respected throughout metamorphosis. Thus, in adults, earlier formed cells are located more rostrally than later ones. In summary, the adult rostral pallium is directly derived from the primitive larval pallium in such a way, the comparison between adult and larval results is justifiable.

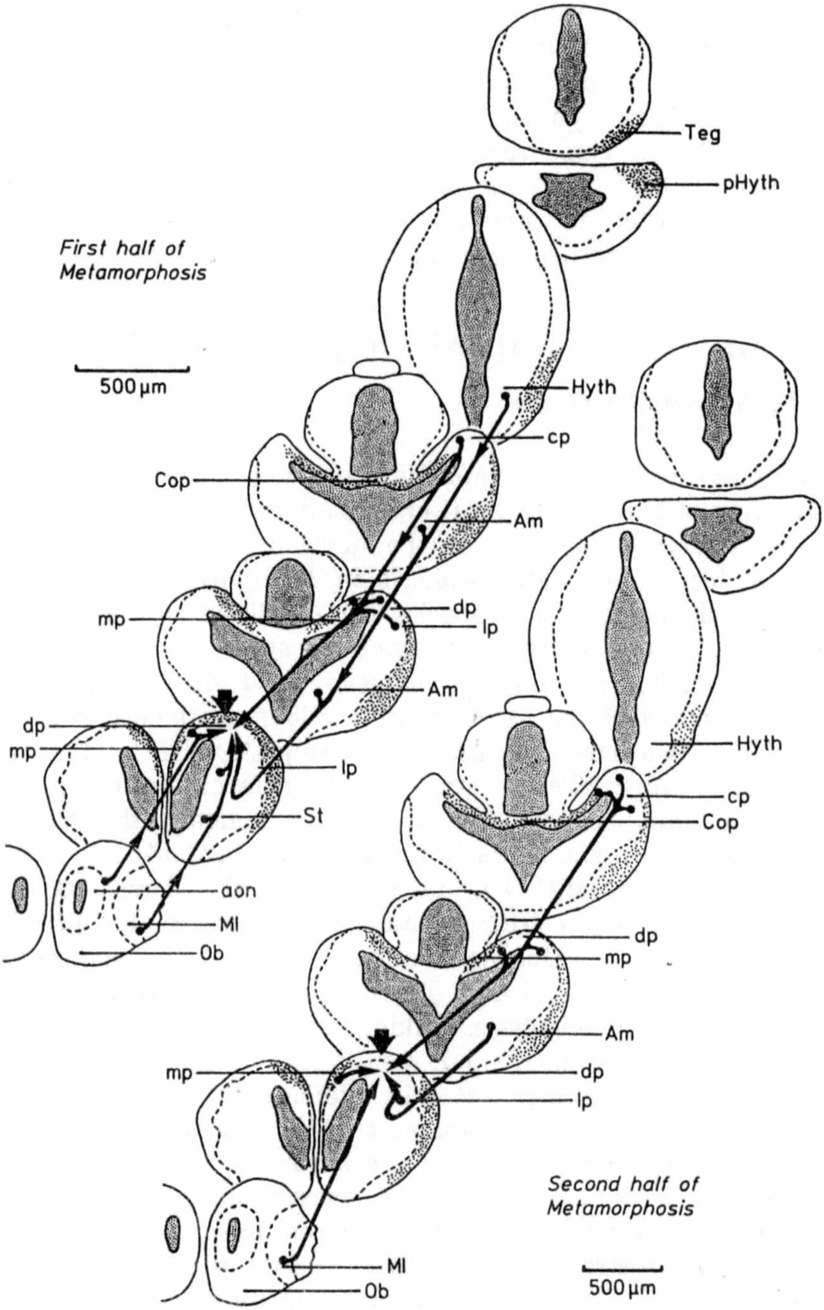

Fig. 2. Series of transversal sections illustrating HRP labeling of the larval dorsal pallium in *Pleurodeles* after HRP injection. Labeled cells are indicated by *solid circles*, fibers by *dotted areas*. The *thick lines* indicate the general pattern of pallial afferents. The *dotted lines* mark the boundary of the periventricular cell masses. *Large arrow*, HRP injection site.

Pallial Afferents

It should be noted that all afferent systems are located in the ipsilateral side during the whole metamorphosis. The following areas displayed labeled cells after injection of HRP:

1. Rostral telencephalon. The distribution of olfactory bulb HRP labeled cells remains the same during the whole metamorphosis, labeled cells being numerous in the olfactory bulb and scarce in the anterior olfactory nucleus (Figs. 3a and 3b).

2. Pallium. During the whole metamorphosis, all pallial areas (i.e. the medial, dorsal and lateral pallium), throughout their rostrocaudal extent, show labeled perikarya (Fig. 3c). Hence, the pallium contains an elaborate association system from the beginning of metamorphosis.

3. Prosencephalic ventro-lateral wall. In the first half of metamorphosis (stage 38 to stage 45), the entire ventro-lateral wall of the hemisphere shows scarce, but well-labeled, neurons. This large area includes the primordium of the striatum (Fig. 3d), the amygdala, the preoptic area and the anterior hypothalamus (Fig. 3E). During the second half of metamorphosis (stage 46 to stage 55), the prosencephalic ventro-lateral wall suffers drastic modifications (Fig. 2). The anterior hypothalamus, the preoptic area and the striatum are now completely free of labeled perikarya, only a few labeled cells are found in the amygdala. Comparing this second set of results with the first one, an impressive conclusion arises: The pallial afferents pattern decreases during metamorphosis.

4. The fibrous network. Many studies underline the capacity of HRP to be transported in both the anterograde and retrograde directions. Labeled fibers may thus be either afferent or efferent. Nevertheless, the lack of labeled perikarya in the contralateral side allows one to conclude that the contralateral fiber network must be exclusively composed of pallial efferents arising in the ipsilateral hemisphere. Anatomically, the fiber network can be divided into two parts: a prominent lateral, ipsilateral one and a smaller medial, bilateral one. The ipsilateral part originates in the bulbar region, travels through the striato-amygdaloid area and extends as far as the anterior ventral tegmentum during the first half of the metamorphosis, while during the second half, its fibers apparently are retracted to end at the rostralmost level of the diencephalon in the anterior hypothalamus (Fig. 2). The bilateral part of the fibrous system extends in a well-delimited dorso-medial portion of the thin medial hemispheric wall, from the rostral pallial area to the level of

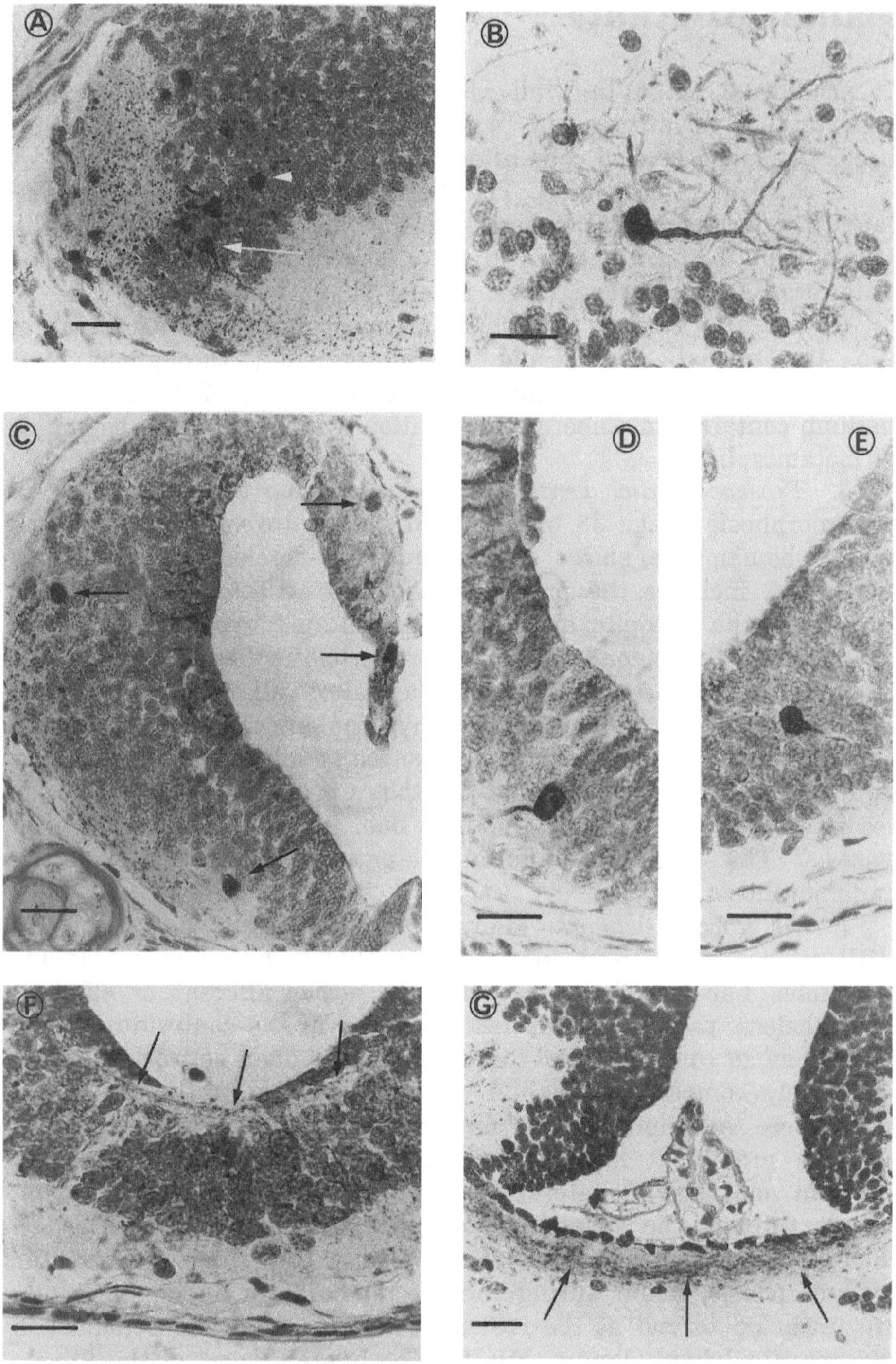
A
B
C
D
E
F
G

the *foramen Monroi*. Ipsilateral and contralateral parts of the medial components are linked via the pallial commissure (Figs. 3f, 3g) where labeled fibers appear as early as in stage 40. No other commissural systems were observed even at the ultimate stages of metamorphosis.

Is the Pallium of Urodela a Compartmentalized Structure?

Harris (1980) confirmed in his study on the salamander that ectopic retinofugal fibers make contacts with central areas. Moreover, he demonstrated the influence of the entrance point of optic fibers into the brain on their pathway and terminal connections. Following an earlier concept (Weiss 1955), Harris concluded that guidance of ingrowing fibers is influenced by external factors specific for each brain region, suggesting the existence of a compartmentalized (or regionalized) organization of the fiber projection pattern in the amphibian brain. Apart from the optic nerve, very few other nerves or associated tracts have so far been investigated to test the generalization of the regionalization hypothesis (or clonal hypothesis, according to Hirose and Jacobson, 1979).

From our results in adult Urodela, *Triturus* and *Pleurodeles*, it appears that the dorsal pallium is composed of at least two distinct regions: (1) The rostral part of the dorsal pallium: After HRP injection in this part, only intratelencephalic systems (including the bulbar afferents) are labeled; (2) The caudal part of the dorsal pallium: Studies on habenular (Clairambault et al. 1986b) and thalamic (Clairambault et al. 1986a; Wicht and Himstedt 1988) projections have shown that only the caudal part of the pallium is connected with the diencephalon. However, to test the regionalization hypothesis using the developing pallium as an experimental model, the location of afferent neurons remains to be

◄────────

Fig. 3 A-G. Larval *Pleurodeles*. **A,B** sagittal sections through the olfactory bulb. **A** stage 41; *arrow* points to mitral cell and *arrowhead* to an HRP filled neuron in the anterior olfactory nucleus. *Bar*: 100 µm. **B** stage 55; mitral cell. *Bar*: 50 µm. **C-G** transverse sections. **C** stage 41; HRP filled cells (*arrows*) in the medial pallium, lateral pallium and striatum. *Bar* = 50 µm. **D** stage 41; HRP filled neuron in the striatum. Bar: 50 µm. **E** stage 43; labeled neuron in the anterior hypothalamus. *Bar*: 50 µm. **F,G** the *commissura pallii* (*arrows*) at stages 43 (**F**) and 55 (**G**). *Bar*: 50 µm.

demonstrated by HRP injections into the caudal pallium alone (work in progress).

Restriction in the Projection Pattern during Metamorphosis

It is well known that restriction (i.e. reduction in length) of a fibrous system can occur in the development of some parts of the brain. One of the best examples is the callosal system: Innocenti et al. (1977) reported that HRP injection into the visual cortex of the young kitten during the first postnatal week resulted in labeling of neurons in areas 17 and 18 of the contralateral side, while in the adult cat the labeling is limited to layers III, IV and VI of the same areas. A similar case of restriction was described in the parietal cortex of the rat (O'Leary et al. 1981).

In *Pleurodeles*, the fibers running in the ipsilateral ventro-lateral wall of the prosencephalon reach the rostral tegmentum at the beginning of metamorphosis and progressively limit their extension to the level of the *area preoptica* during the last part of metamorphosis. This typical case of restriction remains to be explained. Is it a mechanical withdrawal of axons, a degeneration of collaterals or another phenomenon? No available data allows us to give an answer to this puzzling problem.

Can the Developing Pallium Illustrate the "Parcellation Hypothesis"?

The parcellation hypothesis (Ebbesson 1984) proposes that the pattern of interspecific variation of neuronal circuitry among vertebrates is the result of a loss of connections which changes a primitive uniform pattern of overlap into a specialized pattern of isolated fields of projection (segregation, isolation, parcellation (SIP) process). According to Ebbesson (1984, p. 322), the data "suggests that diffuse, relatively undifferentiated systems existed at the beginning of vertebrate evolution... One principle underlying the growth, differentiation... appears to be a process of parcellation that involves the selective loss of connections in daughter systems...".

Because no paleontological data is available in neuroanatomy, we can only conceive of the evolution of the vertebrate brain with the aid of comparative and developmental studies (with Haeckel's theory as the sword of Damocles!). In particular, we think that ontogenetic studies are important if we keep in mind that ontogeny progresses along similar lines in all vertebrates; accordingly, whereas the early stage embryos of all vertebrates are similar, they diverge with maturation and the divergence occurs earlier for more distantly related species. Our belief is that Ebbesson's parcellation hypothesis is more an ontogenetic concept than an evolutionary one.

From our work, we offer two ontogenetic examples which could illustrate the SIP hypothesis as an embryological concept:

1. In a previous paper (Timmel and Clairambault 1988), we reported how the precocious, direct bulbo-habenular tract progressively changes into a polysynaptic bulbo-pallio-habenular pathway during the metamorphosis of *Pleurodeles waltlii*. This case is a typical example of parcellation within one system.

2. An HRP injection into the pallium of Pleurodeles labels neurons in the whole striato-amygdaloid area of the young larva, while in the adult, only a few labeled neurons appear in a small zone along the boundary between the striatum and the anterior amygdala. From this result, we conclude that an early, extensive striato-pallial system is progressively restricted during development to its definitive site of origin.

We do not know whether this anatomical concentration leads to a specific function, as Ebbesson (1984, p. 331) thinks. New experimental probes are required to demonstrate that, during the development of the amphibian telencephalon, generalized larval systems evolve into specialized adult systems.

Acknowledgements. We wish to express our sincere thanks to Mrs. C. Pairault for her technical assistance.

References

Ariens Kappers CU (1908) Eversion and inversion of the dorsolateral wall in different parts of the brain. J Comp Neurol 18:433-436

Ariens Kappers CU, Huber GC, Crosby EC (1936) The comparative anatomy of the nervous system of vertebrates including man. Hafner, New York

Clairambault P (1976) Development of the prosencephalon. In: Llinas V, Precht W (eds) Frog neurobiology. Springer, Berlin, Heidelberg, New York pp 924-945

Clairambault P, Capanna E (1970) Aspetti della neuroistogenesi del pallio telencefalico in due Ranidi. Rend Accad Naz Lincei Serie VIII, 48:727-735

Clairambault P, Picouet MJ, Chanconie M (1973) Morphogenèse du télencéphale des Anoures au cours du développement embyronnaire. J Hirnforsch 14:43-58

Clairambault P, Fasolo A, Pairault C, Levrat-Calviac V (1986a) Pallial connections in Urodela. Neurosci Lett (Suppl) 26:447

Clairambault P, Pairault C, Fasolo A (1986b) Habenular connections in the brain of the newt *Triturus cristatus carnifex L.* J Hirnforsch 27:111-119

Ebbesson SOE (1984) Evolution and ontogeny of neural circuits. Behav Brain Sci 7:321-366

Gallien L, Durocher M (1957) Table chronologique du développement de *Pleurodeles waltlii micha*. Bull Biol Fr Belg 91:97-114

Halpern M (1972) Some connections of the telencephalon of the frog *Rana pipiens*. Brain Behav Evol 6:42-68

Harris WA (1980) Regions of the brain influencing the projection of developing optic tracts in the salamander. J Comp Neurol 194:319-333

Herrick CJ (1948) The brain of the tiger salamander. University of Chicago Press, Chicago.

Hirose G, Jacobson M (1979) Clonal organization of the central nervous system. I. Clones stemming from individual blastomeres of the 16-cell and earlier stages. Dev Biol 71:191-202

Hsu SN, Raine L, Fanger H (1981) Use of avidin-biotin peroxidase complex (ABC) in immunoperoxidase techniques: a comparison between ABC and unlabeled antibody (PAP) procedures. J Histochem Cytochem 29:577-580

Innocenti GM, Fiore L, Caminiti R (1977) Exuberant projection into the corpus callosum from the visual cortex of newborn cats. Neurosci Lett 4:237-242

Northcutt RG (1981) Evolution of the telencephalon in nonmammals. Annu Rev Neurosci 4:301-350

Northcutt RG, Royce GJ (1975) Olfactory bulb projections in the bullfrog *Rana catesbeiana*. J Comp Neurol 145:251-267

Northcutt RG, Kicliter E (1980) Organization of the Amphibian telencephalon. In: Ebbesson SOE (ed) Comparative neurology of the telencephalon. Plenum, New York, pp 203-255

O'Leary DDM, Stanfield BB, Cowan WM (1981) Evidence that the early postnatal restriction of the cells of origin of the callosal projection is due to the elimination of axonal collaterals rather than to the death of neurons. Dev Brain Res 1:607-617

O'Leary DDM, Gerfen CR, Cowan WM (1983) The development and restriction of the ipsilateral retinofugal projection in the chick. Dev Brain Res 10:93-109

Reiner A, Northcutt RG (1987) An immunohistochemical study of the telencephalon of the African lungfish *Protopterus annectens*. J Comp Neurol 256:463-481

Shumway W (1940) Normal stages in the development of *Rana pipiens*. Anat Rec 78:139-147

Timmel JF, Clairambault P (1988) Premières données sur la mise en place des connexions habénulaires au cours de développement chez *Pleurodeles waltii* (Amphibia, Urodela). C R Acad Sci Paris, Serie III, 306:507-510

Weiss P (1955) The nervous system. In: Willier BH, Weiss P, Hamburger V (eds) Analysis of development. Saunders, Philadelphia, pp 346-401

Wicht H, Himstedt W (1988) Topologic and connectional analysis of the dorsal thalamus of *Triturus alpestris* (Amphibia, Urodela, Salamandridae) J Comp Neurol 267:545-561

Wilczynski W, Northcutt RG (1983) Connections of the bullfrog striatum. Efferent projections. J Comp Neurol 214:333-343

Abbreviations

Am, amygdala; **aon**, anterior olfactory nucleus; **aonmp**, anterior olfactory nucleus - medial part; **aonvp**, anterior olfactory nucleus - ventral part; **Ap**, area preoptica; **Coh**, commissura habenularis; **Cop**, commissura palii; **cp**, caudal pallium; **dp**, dorsal pallium; **fM**, foramen Monroi; **Gl**, granular cell layer of the olfactory bulb; **gll**, glomerular layer of the olfactory bulb; **Hyth**, hypothalamus; **lp**, lateral pallium; **mp**, medial pallium; **Ml**, mitral cell layer of the olfactory bulb; **Ob**, olfactory bulb; **On**, olfactory nerve; **pHyth**, posterior hypothalamus; **Sm**, stria medullaris; **Sp**, septum; **St**, striatum; **Teg**, tegmentum; **Th**, thalamus.

Brain Stem Projections to the Telencephalon in Two Species of Amphibians, *Triturus alpestris* (Urodela) and *Ichthyophis kohtaoensis* (Gymnophiona)

Helmut Wicht and Werner Himstedt
Institut für Zoologie der TH Darmstadt
Schnittspahnstraße 3, D-6100 Darmstadt, FRG

The majority of ascending inputs to the vertebrate telencephalon is relayed in the dorsal thalamus. Inputs arising from the brain stem are far less spectacular in quantitative terms, nevertheless they are of high functional importance. In mammals, the integrity of the (dopaminergic) nigro-striatal system is known to be a prerequisite for the feedback control of motor activity, and the (catecholaminergic) coeruleo-telencephalic and (serotoninergic) raphe-telencephalic systems play an important role in the organization of the different phases of sleep and wakefulness. But even sensory inputs reach the telencephalon via direct channels from the brain stem. For example, in birds, the principal sensory nucleus of the trigeminal nerve, which receives tactile information from the beak via the trigeminal and hypoglossal nerve (Dubbeldam 1980), gives rise to the tractus quinto-frontalis (Wallenberg 1903). This tract projects directly to the rostral part of the telencephalic lobus parolfactorius, namely, to the nucleus basalis.

There are few experimental observations on the organization of the telencephalic afferents from the brain stem in amphibians. Among the three orders of recent amphibia (Anura, Urodela, and Gymnophiona or Apoda) only the anurans have received attention with regard to the organization of these projection systems (Kicliter 1979; Northcutt and Kicliter 1980; Wilczynski and Northcutt 1983 a,b). To our knowledge, there are no experimental studies on the ascending projections from the brain stem either in urodeles or in gymnophiones. However, the knowledge of the organization of these systems among amphibia is of some interest in phylogenetic considerations: since amphibia are regarded as the sister group of amniotic vertebrates (Ax 1984), their brain characters play an important role in the attempt to establish homologies and to

Experimental Brain Research Series 19

recognize plesiomorphic vs. apomorphic characters among amniotes.

During a series of experiments which aimed at the clarification of the thalamo-telencephalic connectivity in urodeles (Wicht and Himstedt 1988) and gymnophiones (Wicht et al., in preparation), several brain stem projections to the telencephalon of these amphibians were detected. Our data support the idea of a relative stability of some of these projections throughout the vertebrate phylogeny and, in comparison to the other hodological and histochemical studies on the amphibian brain, they allow the identification of some cell groups - especially in the urodele brain - which have not been recognized so far.

Material and Methods

Horseradish peroxidase (HRP) was injected into the forebrains of 19 *Triturus alpestris* (Amphibia, Urodela, Salamandridae) and of 6 *Ichthyophis kohtaoensis* (Amphibia, Gymnophiona, Ichthyophidae). In most of the cases either the medial or the lateral forebrain bundle or both were lesioned unilaterally at middle or posterior telencephalic levels and HRP was applied to the cut fiber bundle. In some cases diluted HRP was injected iontophoretically into the striatum of *Triturus*. Most of the experiments in *Triturus* were carried out under "in vivo" conditions, but some of the data for *Triturus* and all of the data for *Ichthyophis* were obtained using a slightly modified version of Ekström's (1985) "in vitro" technique (for details see Wicht and Himstedt 1988).

The brain slices (40 µm) were reacted for HRP using the catalytic properties of the transported enzyme itself; diaminobenzidine (LaVail and LaVail 1972) was used as chromogen. In several cases the reaction product was intensified by addition of nickel and cobalt salts to the incubation medium (Adams 1981).

Results

Terminology

The terminology for the telencephalon of both *Triturus* and *Ichthyophis* is adopted from Northcutt and Kicliter (1980), the other terms are taken from Herrick's (1948) description of the brain of the tiger salamander and from Kuhlenbeck's (1922) monograph on the brain of gymnophiones. It must, however, be mentioned that the identification of some areas in the gymnophionan brain (especially in the mes- and metencephalon) is somewhat tentative. Major landmarks (like the cerebellum) are absent or currently unrecognized (see Kuhlenbeck 1975, 1978), and the descriptive anatomy is complicated by the pronounced flexures of the brain. In order to make possible a critical evaluation of our identification of brain stem areas in gymnophiones (since their brains may be unfamiliar even to specialists in comparative anatomy) we shall start with a short description of the general morphology of the brain of *Ichtyophis*. A more detailed analysis, especially of the diencephalon, is in preparation.

Ichthyophis kohtaoensis

A lateral view of the brain can be seen on the right side of Fig. 1. The telencephalon dominates the brain in this view, especially the poli posteriores (pol.po.) are very large and project backwards to mid-mesencephalic levels, covering almost the entire diencephalon. Only the most ventral parts of the diencephalon are visible: the large hypophysis and the ventral part of the hypothalamus with the entry of the thin optic nerve. The brain displays a sharp bend in the "pontine flexure" (arrow in Fig. 1), which deeply affects the morphology of the transitional zone between mes-, met-, and myelencephalon. As a result of this bend, the anterior part of the fossa rhomboidea is displaced rostrally beneath the mesencephalon, and the normal rostro-caudal sequence of the various anatomical subdivisions of this brain area reverts to a dorsoventral one (Fig. 1A).

Fig. 1C shows the extent of a telencephalic injection of HRP, which involved both medial and lateral forebrain bundle. The positions of neurons which were labeled by retrograde transport of HRP are depicted by black dots in A and B. In the floor of the anterior fossa rhomboidea there is a group of labeled cells close to the midline (Fig. 1A). The midline is marked by the sulcus medianus internus (s.med.int.). Due to the extreme bending of the brain in the "pontine flexure" the anterior part of this sulcus does not run in a rostro-caudal, but rather in a dorsoventral direction, thus, in a transverse plane of section, it looks like a slit rather than a groove. Another group of cells can be observed in the transitional area between tegmentum isthmi (teg.is.) and the tegmentum trigemini (teg.V), ipsilateral to the injected side. In the rostral mesencephalon, there are cells in the "peduncular area" (ped.), i.e., the part of the tegmentum which surrounds the tuberculum posterius (Fig. 1B). The cells form a column which extends from the tip of the tuberculum posterius to the exit of the third nerve. Still another group of neurons which project to the telencephalon is visible in Fig. 1B: it is located in the dorsal part of the infundibular hypothalamus, ipsilateral to the injected side.

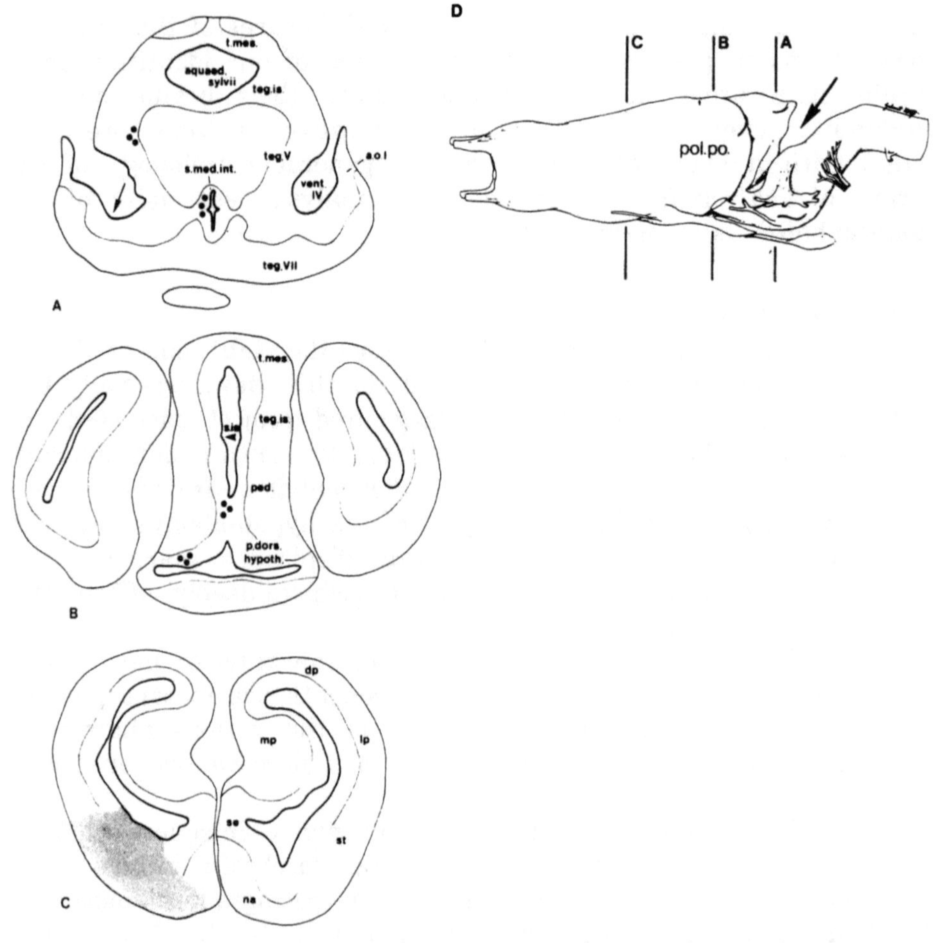

Fig. 1 A-C. Transverse sections through the brain of *Ichthyophis kohtaoensis*. *Black dots* mark the positions of neurons labeled by telencephalic injection of HRP (*shaded area in C*). *Small arrow* in **A** points towards the sulcus limitans of His. **D** Lateral view of the brain of *Ichthyophis* indicating the levels of the sections shown in **A**, **B** and **C**. The *arrow* points into the cavity of the "pontine flexure". *Aquaed.sylvii*, aquaeductus sylvii; *a.o.l.*, area octavo-lateralis; *cb.*, cerebellum; *com.tub.po.*, commissura tuberculi posterioris; *dp*, dorsal pallium; *III*, nervus oculomotorius; *l.l.cb.*, lobus lateralis cerebelli; *lp*, lateral pallium; *mp*, medial pallium; *na*, nucleus accumbens; *p.dors.hypoth.*, pars dorsalis hypothalami; *ped.*, "peduncle"; *pol.po.*, polus posterior; *se*, septum; *s.is.*, sulcus isthmi; *s.med.int.*, sulcus medianus internus; *st*, striatum; *teg.dors.*, tegmentum dorsale; *teg.is.*, tegmentum isthmi; *teg.V.*, tegmentum trigemini; *teg.VII*, tegmentum nervi facialis; *t.mes.*, tectum mesencephali; *V*, nervus trigeminus; *vent.IV*, ventriculus quartus.

Triturus alpestris

The results obtained in *Triturus* are depicted in Fig. 2. Black dots in Figs. 2A-D mark the position of retrogradely labeled neurons after basal telencephalic injections of HRP (Fig. 2E).

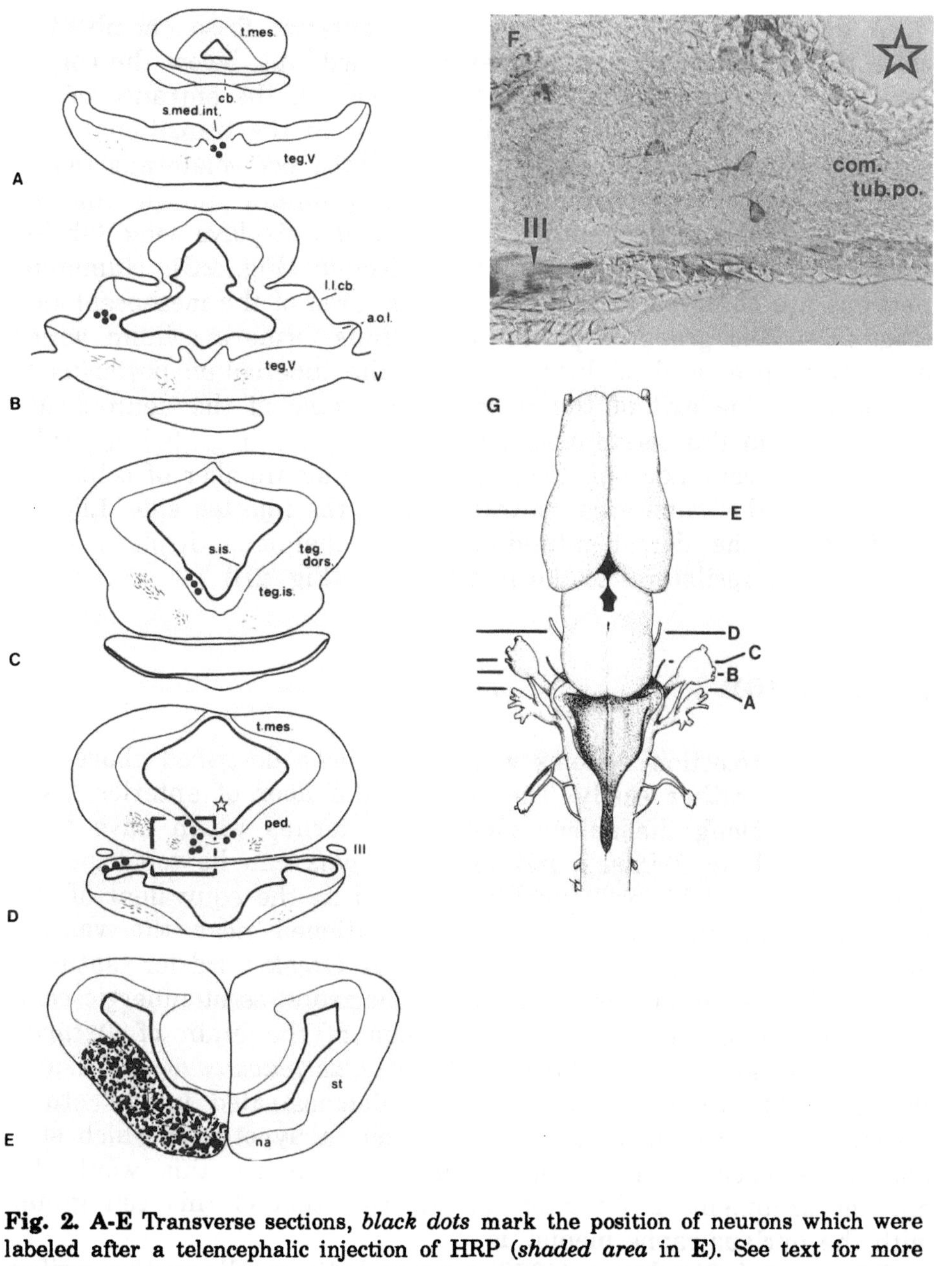

Fig. 2. A-E Transverse sections, *black dots* mark the position of neurons which were labeled after a telencephalic injection of HRP (*shaded area* in **E**). See text for more detailed description. **F** Ventrally displaced neurons in the "peduncular area" (see *insert* in **D**). **G** Dorsal view of the brain of *Triturus alpestris*, indicating the planes of the transverse sections shown in **A-E**. For abbreviations see Fig. 1

A group of cells in the floor of the anterior fossa rhomboidea, close to the sulcus medianus internus (s.med.int.), below the corpus cerebelli (c.cb.) is seen in Fig. 2A. At levels of the entrance of the trigeminal nerve (V) there are labeled cells in the transitional area between lobus lateralis cerebelli (l.l.cb.), area octavolateralis (a.o.l.) and the tegmentum trigemini (teg. V), ipsilateral to the injected side (Fig. 2B). The tegmentum isthmi (teg.is.) displays some labeled cells on the side ipsilateral to the injection (Fig. 2C). Numerous neurons are observed in the "peduncular" area of the mesencephalon (Fig. 2D). This group of projection neurons forms a column which extends from a level slightly rostral to the tuberculum posterius to the level of the exit of the third nerve. Some of the neurons are detached from the periventricular gray and have migrated ventrally into the adjacent neuropil (Fig. 2F). A smaller number of cells was observed in the same area contralateral to the injected side. Like in *Ichthyophis*, the dorsal infundibular hypothalamus displays some labeled cells, ipsilateral to the injected side (Fig. 2D).

Discussion

Two of the projection systems which have been described above can be identified rather easily. The cells in the floor of anterior fossa rhomboidea along the sulcus medianus internus which have been observed both in *Triturus* and in *Ichthyophis* can be topologically, hodologically and histochemically identified as the equivalent of the nucleus raphe medianus. They are positioned near the ventral midline suture of the brain ("raphe" is the Greek word for "suture") and display ascending connections. Furthermore, serotoninergic cells have been described in the same region of the brain of *Triturus cristatus* (Fasolo et al. 1986) and of *Necturus maculosus* (Dubé and Parent 1982). Provided that the cells demonstrated in our study actually use serotonin as neurotransmitter (a hypothesis which still has to be checked in a double-label experiment), this would be another argument in favor of the identification of this cell group with the nucleus raphe medianus.

Ronan and Northcutt (1979) reported that cells with similar position and connectional properties project to the medial pallium of the telencephalon. However, in our series of injections of HRP into the medial pallium of *Triturus*, we did not observe the latter projection (unpublished data).

Cell groups comparable to the nucleus raphe medianus have been identified in a variety of fishes: elasmobranchs (Fiebig and

Bleckmann 1988), *Polypterus* (Northcutt 1981), and actinopterygians (Murakami et al. 1983). Data on agnathans and lungfishes are lacking. A nucleus raphe medianus is present in reptiles (hodological data: Bruce and Butler 1984; Ouimed et al. 1985; ten Donkelaar and de Boer-van Huizen 1988; immunohistochemical data: Smeets 1988; Smeets and Steinbusch 1988), it is also present in birds (Miceli and Réperant 1985; Wild 1987), and, of course, in mammals (e.g. Vertes 1988; for a detailed immunohistochemical study see Hornung and Fritschy 1988). Thus the presence of a serotoninergic raphe-telencephalic system seems to be a feature of ancient gnathostomes, however, with considerable variations in detail. For example, various subdivisions of the raphe area are recognizable in sauropsids and mammals; up to seven different raphe nuclei can be observed in the marmoset (Hornung and Fritschy 1988), and at least three raphe nuclei have been described in a lizard (ten Donkelaar and de Boer-van Huizen 1988). The amphibian raphe is much less differentiated, especially the lateral subdivisions of the raphe complex (nucleus lateralis raphe superioris), and the caudal subdivisions (nucleus raphe inferioris) seem to be absent.

The second cell group which can be identified rather easily is the group located in the "peduncular area" of both *Triturus* and *Ichthyophis*. The term "peduncle" was used by Herrick (1948) to denote that part of the tegmentum which surrounds the tuberculum posterius. Since nothing comparable to the mammalian or human pedunculus cerebri is present in the amphibian brain (see Herrick 1948, p. 21) we prefer to replace that expression by the more neutral term "posterior tubercular area".

Topological, hodological and histochemical characteristics of the cell group in this area in amphibians suggest that it is the equivalent of the substantia nigra, pars compacta and/or the ventral tegmental area of other vertebrates. Like those, the cell group is located in the anterior ventral tegmentum of the mesencephalon, at levels of the exit of the third nerve. Like in mammals and sauropsids the cell group is situated in close vicinity to another typical structure of this area: the nucleus ruber, which recently has been identified at least in urodeles (Naujoks-Manteuffel and Manteuffel 1988). In an immunohistochemical investigation of the urodele brain (Franzoni et al. 1986), it has been shown that tyrosine-hydroxylase positive neurons are present in the posterior tubercular area, in a position comparable to the cell group we have shown to possess ascending connections. Already in 1982 Dubé and Parent observed a catecholaminergic fiber system arising from cells in the same region of the brain of *Necturus*, which ends in the striatum. A similar projection system has been described for anurans (from cells adjacent to the nucleus tuberculi posterioris; Wilczynski and Northcutt 1983a), and it has been argued that this system corresponds to the substantia nigra, pars compacta of amniotic vertebrates (Wilczynski and Northcutt 1983b).

Like the raphe-telencephalic system, the nigro-striatal system appears to be a feature of ancient gnathostomes. It is present in elasmobranchs (hodological data: Fiebig and Bleckmann 1988; immunohistochemical data: Meredith and Smeets 1987), in *Polypterus* (Northcutt 1981), and in reptiles (ten Donkelaar and de Boer-van Huizen 1988). In birds, however, the areas in the mesencephalon which give rise to a catecholaminergic innervation of the telencephalon bear other names: area ventralis of Tsai and nucleus tegmento-pedunculo-pontinus (Kitt and Brauth 1986). Actinopterygians display a very complex organization of their corresponding projection systems: it has been argued by Braford and Northcutt (1983) and by Northcutt and Wullimann (1988) that the entire "preglomerular complex" (a large set of nuclei located in the basal diencephalon including nucleus praethalamicus and praeglomerulosus) is a derivate of the posterior tubercular area. In advanced actinopterygian fishes, these nuclei are one of the major sources of input to the telencephalon. Again, as compared with other vertebrates, the organization of the amphibian nigro-striatal system appears much simpler.

It is so far more difficult to identify the other cell groups in the urodele and gymnophionan brain, which display ascending projections to the telencephalon. The cells in the transitional area between tegmentum trigemini, area octavo-lateralis, and lobus lateralis cerebelli in *Triturus* (Fig. 2B) and their counterpart in *Ichthyophys* (Fig. 1A) can be interpreted in different ways. Cells projecting to the posterior (amygdalar) telencephalon have been observed in a roughly similar position in anurans, namely, in the nucleus cerebelli (Kicliter 1979). It has been argued that the telencephalopetal projection, which arises from these cells, might represent a homologue of the tractus quinto-frontalis of birds (Northcutt and Kicliter 1980). If this hypothesis was also valid for urodeles, the cells should receive direct inputs from the maxillo-mandibular branches of the trigeminal nerve or should be topographically continuous with the principal sensory trigeminal nucleus. However, a recent investigation in the urodele *Pleurodeles* (Gonzales and Muñoz 1988) showed that the cells under consideration probably do not receive direct trigeminal inputs. They also do not seem to be continuous with the principal sensory trigeminal nucleus (see Fig. 4 of Gonzales and Muñoz 1988).

Another possibility is to identify the cells with the locus coeruleus of other vertebrates, the more so since catecholaminergic neurons have been observed in similar location in the brains of *Triturus cristatus* (Franzoni et al. 1986) and of *Necturus maculosus* (Dubé and Parent 1982). However, both Dubé and Parent and Franzoni et al. denied the existence of an ascending catecholaminergic pathway arising from this area. They stated that the main catecholaminergic efferents of this field are directed towards the cerebellum and the spinal cord. Although we have found cells projecting to the telencephalon in this area, it would be premature to propose, on the basis of this finding alone, that urodeles and gymnophiones actually possess a homologue of the amniote locus coeruleus. Clearly, a double-lable experiment is needed in order to check whether the projection neurons with ascending projections we have observed in this area actually use catecholamines as neurotransmitters.

The group of cells with ascending projections in the tegmentum isthmi occurred only in *Triturus*. It is located in a position which corresponds to the nucleus ventralis tegmentalis of Naujoks-Manteuffel and Manteuffel (1988), a cell group which previously was known to have mainly descending connections to the medulla oblongata. A nucleus roughly comparable in position and also possessing sparse ascending projections to the telencephalon has been identified in anurans, namely, the superficial isthmal reticular nucleus (Wilczynski and Northcutt, 1983a). This nucleus in the anuran brain additionally displays projections to the midbrain roof (Wilczynski 1981) and receives a strong input from the striatum. Thus it is a possible candidate for a homologue to the substantia nigra pars reticularis, as Wilczynski and Northcutt (1983b) have pointed out. This homology, however, cannot be transferred to the nucleus ventralis tegmentalis of urodeles since there is no projection from this area to the midbrain roof. The main tegmental efferents to the tectum arise from cells located dorsal to the nucleus ventralis tegmentalis (Finkenstädt et al. 1983; Rettig 1988). All in all, it is presently not possible to identify this cell group with ascending connections in a comparative way of reasoning.

Finally, we observed cells in the dorsolateral infundibular hypothalamus both in *Triturus* and *Ichthyophis*, close to the lateral recess of the infundibulum. The same region contains serotoninergic and catecholaminergic cells (Fasolo et al. 1986; Franzoni et al. 1986). A comparable cell group has not been identified in anurans

(Kicliter 1979; Wilczynski and Northcutt 1983a), nor in teleosts, *Polypterus* or birds. However, a telencephalic projection from the posterior hypothalamus has been reported for elasmobranchs (Fiebig and Bleckmann 1988), for reptiles (from the corpus mammillare, ten Donkelaar and de Boer-van Huizen 1988), and for mammals (from the nucleus supramammillaris, Vertes 1988). The patchy distribution of this character within the different groups of vertebrates does not yet allow us to draw final phylogenetic conclusions. An investigation of the afferents of the telencephalon in agnathans (which is still lacking) should help to clarify whether presence or absence of a posterior hypothalamo-telencephalic projection system is the plesiomorphic or apomorphic condition for gnathostomes.

Concluding Remarks

Urodeles and gymnophiones possess a set of brain stem projections upon the telencephalon which, in part, can readily be compared with those present in other gnathostomes. Raphe-telencephalic, nigro-telencephalic, and, possibly, coeruleo-telencephalic systems have been found. In addition, a hypothalamo-telencephalic connection arises from the posterior hypothalamus, a projection which occurs only in a few other vertebrate groups. Whereas there are no "unusual" sources of brain stem inputs to the telencephalon, some ascending systems which have been described in other vertebrates are absent: there are no signs of a toro-telencephalon projection, which has been described, e.g., for *Polypterus* (Northcutt 1981) and a lizard (ten Donkelaar and de Boer-van Huizen 1988). Likewise, a projection from the so-called substantia grisea centralis mesencephali (central gray), which is present in elasmobranchs (Fiebig and Bleckmann 1988) in a lizard (ten Donkelaar and de Boer-van Huizen 1988), in birds (Miceli and Réperant 1985), and in mammals, was not detected in amphibians.

Thus, the amphibian brain has its own, unique set of characters, and one should be careful when speaking about the amphibian - and especially the urodele - brain as a model, a "pattern or template" or "a standard of reference in the study of all other vertebrate brains." (Herrick 1948, p. 17). In the realm of pure morphology it may indeed be justified to approach the amphibian brain with this idea in mind (e.g. Kuhlenbeck 1978, p. 426). In evolutionary morphology, however, one should not forget that "the living species of the groups of vertebrates represent the upper tips of the widely branched vertebrate phylum and thus are not to be derived from each other. In the same way it is impossible to derive the brain structures of one of these groups from those of another" (Holmgren 1922, p. 345).

Acknowledgements. We are obliged to Drs.H.Bleckmann and E.Fiebig for use of the entire text of the contribution they presented in Göttingen in 1988, and to Prof.Dr.A.-Fasolo for an inspiring visit in Torino. This study was supported by a scholarship of the Studienstiftung des Deutschen Volkes to H.W. and a grant from the Deutsche Forschungsgemeinschaft (DFG Hi 306/2-1).

References

Adams JC (1981) Heavy metal intensification of DAB-based HRP-reaction product. J Histochem Cytochem 29:775

Ax P (1984) Das phylogenetische System. Fischer, Stuttgart

Braford MA, Northcutt RG (1983) Organization of the diencephalon and pretectum of the ray-finned fishes. In: Davis RE, Northcutt RG (eds) Fish neurobiology, Vol.3. University of Michigan Press, Ann Arbor, pp 117-163

Bruce LL, Butler AB (1984) Telencephalic connections in lizards. I. Projections to cortex. J Comp Neurol 229:585-601

Dubbeldam JL (1980) Studies on the somatotopy of the trigeminal system in the mallard (*Anas platyrhynchos L.*). I. Morphology of the principal sensory nucleus. J Comp Neurol 191:557-571

Dubé L, Parent A (1982) The organization of the monoamine-containing neurons in the brain of the salamander, *Necturus maculosus*. J Comp Neurol 211:21-30

Ekström P (1985) Anterograde and retrograde filling of central neuronal systems with horse-radish peroxidase under in vitro-conditions. J Neurosci Methods 15:21-35

Fasolo A, Franzoni MF, Gaudino G, Steinbusch HWM (1986) The organization of the serotonin-immunoreactive neuronal systems in the brain of the crested newt, *Triturus cristatus carnifex Laur.* Cell Tissue Res 243:239-247

Fiebig E, Bleckmann H (1988) Telencephalic afferents in the thornback guitarfish, *Platyrhiniodis triseriata* (Elasmobranchii). A WGA-HRP study. In: Elsner N, Barth FG (eds) Proceedings of the 16th Göttingen neurobiology conference, Thieme, Stuttgart, p 304

Finkenstädt T, Ebbesson SOE, Ewert J-P (1983) Projections to the midbrain tectum in *Salamandra salamandra*. Cell Tissue Res 234:39-55

Franzoni MF, Thibault J, Fasolo A, Martinoli MG, Scaranari F, Calas A (1986) Organization of tyrosine-hydroxilase immunopositive neurons in the brain of the crested newt, *Triturus cristatus carnifex*. J Comp Neurol 251:121-134

Gonzales A, Muñoz M (1988) Central distribution of the efferent cells and the primary afferent fibers of the trigeminal nerve in *Pleurodeles waltlii* (Amphibia, Urodela). J Comp Neurol 270:517-527

Herrick CJ (1948) The brain of the tiger salamander. University of Chicago Press, Chicago

Holmgren N (1922) Points of view concerning forebrain morphology in lower vertebrates. J Comp Neurol 34:390-459

Hornung JP, Fritschy JM (1988) Serotoninergic system in the brain stem of the marmoset: A combined immunocytochemical and three-dimensional reconstruction study. J Comp Neurol 270:471-487

Kicliter E (1979) Some telencephalic connections in the frog, *Rana pipiens*. J Comp Neurol 185:75-86

Kitt CA, Brauth SE (1986) Telencephalic projections from midbrain and isthmal cell groups in the pigeon. II. The nigral complex. J Comp Neurol 247:92-110

Kuhlenbeck H (1922) Zur Morphologie des Gymnophionengehirnes. Jena Z Naturwiss 58:453-484

Kuhlenbeck H (1975) The central nervous system of vertebrates, Vol. 4. Karger, Basel

Kuhlenbeck H (1978) The central nervous system of vertebrates, Vol. 5/2. Karger, Basel

LaVail JH, LaVail MM (1972) Retrograde axonal transport in the central nervous system. Science 176:1416-1417

Meredith GE, Smeets WJAJ (1987) Immunocytochemical analysis of the dopamine-system in the forebrain and midbrain of *Raja radiata*: Evidence for a substantia nigra and a ventral tegmental area in cartilaginous fish. J Comp Neurol 265:530-548

Miceli D, Réperant J (1985) Telencephalic afferent projections from the diencephalon and the brain stem in the pigeon. A retrograde multiple-label fluorescent study. Exp Biol 44:71-99

Murakami T, Morita J, Ito H (1983) Extrinsic and intrinsic fiber connections of the telencephalon in a teleost, *Sebasticus marmoratus*. J Comp Neurol 216:115-131

Naujoks-Manteuffel C, Manteuffel G (1988) Origins of descending projections to the medulla oblongata and rostral medulla spinalis in the urodele *Salamandra salamandra* (Amphibia). J Comp Neurol 273:187-206

Northcutt RG (1981) Localization of neurons afferent to the telencephalon in a primitive bony fish, *Polypterus palmas*. Neurosci Lett 22:219-222

Northcutt RG, Kicliter E (1980) Organization of the amphibian telencephalon. In: Ebbesson SOE (ed) Comparative neurology of the telencephalon. Plenum, New York, pp 203-255

Northcutt RG, Wullimann MF (1988) The visual system in teleost fishes: morphological patterns and trends. In: Altema J, Popper AN, Tavolga WN (eds) Sensory biology of aquatic animals. Springer, Berlin, Heidelberg, New York

Ouimed CC, Patrick RL, Ebner FF (1985) The projection of three extrathalamic cell groups to the cerebral cortex of the turtle *Pseudemys*. J Comp Neurol 237:77-84

Rettig G (1988) Connections of the tectum opticum in two urodeles, *Salamandra salamandra* and *Bolithoglossa subpalmata*, with special reference to the nucleus isthmi. J Hirnforsch 29:5-16

Ronan MC, Northcutt RG (1979) Afferent and efferent connections of the bullfrog medial pallium. Neurosci Abstr 5:146

Smeets WJAJ (1988) The monoaminergic system of reptiles investigated with antibodies against serotonin, dopamine and noradrenaline. In: Schwerdtfeger WK, Smeets WJAJ (eds) The forebrain of reptiles. Karger, Basel, pp 97-109

Smeets WJAJ, Steinbusch HWM (1988) Distribution of serotonin immunoreactivity in the forebrain and midbrain of the lizard *Gekko gekko*. J Comp Neurol 271:419-434 ten Donkelaar HJ, de Boer-van Huizen R (1988) Brain stem afferents to the anterior dorsal ventricular ridge in a lizard, *Varanus exanthematicus*. Anat Embryol 177:465-475

Vertes RP (1988) Brain stem afferents to the basal forebrain in the rat. Neuroscience 24:907-935

Wallenberg A (1903) Der Ursprung des Tractus isthmo-striatus (oder bulbo-striatus) der Taube. Neurol Zentralbl 22:98-101

Wicht H, Himstedt W (1988) Topologic and connectional analysis of the dorsal thalamus of *Triturus alpestris* (Amphibia, Urodela, Salamandridae). J Comp Neurol 267:545-561

Wilczynski W (1981) Afferents to the midbrain auditory center in the bullfrog, *Rana catesbeiana*. J Comp Neurol 198:421-433

Wilczynski W, Northcutt RG (1983a) Connections of the bullfrog striatum: afferent organization. J Comp Neurol 214:321-332

Wilczynski W, Northcutt RG (1983b) Connections of the bullfrog striatum: efferent projections. J Comp Neurol 214:333-343

Wild JM (1987) Thalamic projection to the paleostriatum and the neostriatum in the pigeon *(Columba livia)*. Neuroscience 20:305-327

Organization of the Basal Telencephalon in Urodela

A. Fasolo, M. Sassoè Pognetto, A. Battaglia,
M.F. Franzoni*, P. Clairambault**, A. Contestabile***
Dip. Biologia Animale, Università di Torino, Italy
* Dip. Biologia Cellulare, Università della Calabria, Italy
** Lab. Anatomie Comparée, Université Paris VII, France
*** Lab. Biologia Animale, Università di Catania, Italy

On the basis of neuronal organization and connection pattern, the brain of Urodela was considered as a simplified template of the vertebrate brain (Herrick 1948). However, both true primitivity and involution have been discussed as the cause of such a simple organization.

Recently, new powerful techniques, such as axonal transport and immunohistochemistry, have given some clues, helped to draw homologies and helped to understand to what extent the urodele brain can be considered the *Bauplan* of the tetrapod one.

Previous studies had shown that cell groups in the basal telencephalon could be defined on the basis of multiple parameters (including topological, hodological and neurochemical features; Reiner et al. 1984). Furthermore, such specific cell populations have apparently been conserved during evolution of the basal ganglia in amniotes (Reiner et al. 1984; Parent 1986). Yet, the striatum of amphibians, at least that of Anura, seems to be organized according to a pattern that is unique among tetrapods (Parent 1986).

We thought, hence, that a study of the simple periventricular striatal gray of a urodele, the crested newt *Triturus carnifex*, might be of interest in order to compare its morphofunctional organization with that shown in other tetrapods.

Neurochemical Organization of Newt Striatum

New data on the neurochemical and immunohistochemical features of the striatum in the newt revealed some remarkable similarities to the striatal organization in Sauropsida.

Experimental Brain Research Series 19

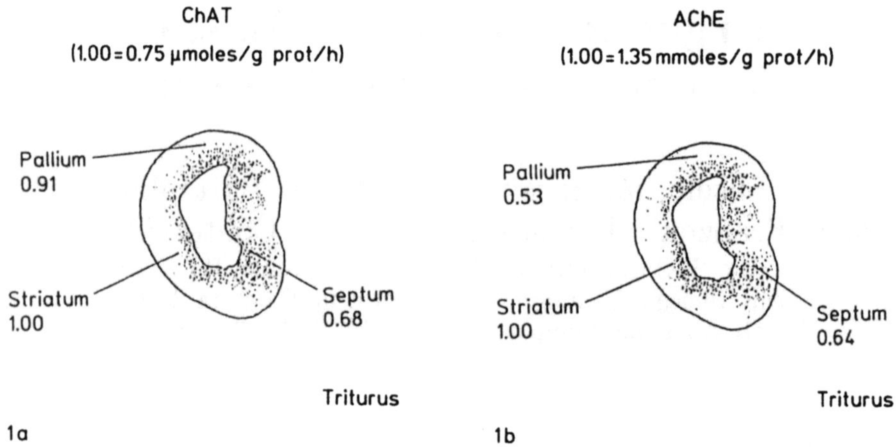

Fig. 1 a,b. Diagrams showing the distribution of choline acetyltransferase (ChAT) (Fig. **1a**) and acetylcholinesterase (AChE) (Fig. **1b**) in a transverse section of the newt telencephalic hemisphere. The results are expressed by attributing the value 1.00 to the brain structure with the highest level of the marker and the corresponding percentage value to the other structures. *h*, hour. For details on the quantitative estimations see Bissoli et al. (1988).

Cholinergic neurotransmission. Investigated by sensitive microchemical methods (see Contestabile and Bissoli, this volume), the striatum of the newt exhibits a fairly high content of a specific cholinergic marker, the enzyme choline acetyltransferase (ChAT; Fig. 1a). Another, yet less specific, cholinergic marker is the enzyme acetyl cholinesterase (AChE). AChE reaches higher levels in the striatum than in the other telencephalic regions (Fig. 1b), and accordingly, heavy histochemical staining is achieved in the striatal neuropil (Fig. 2).

Substance P. Substance P (SP)-like immunoreactivity is present in the striatum. SP-positive cell bodies are stained in dorsal and ventral locations and in the amygdaloid region (Fig. 3). A considerable amount of labeled fiber systems are encountered, including several components of the lateral forebrain bundle. Present data are in agreement with a previously published paper on the newt (Taban and Cathieni 1983) which reported SP-like immunoreactive fiber bundles originating from the ventral striatum and taking the form of pipe-shaped structures. Such data differ from observations on the striatum of the frog, *Rana catesbeiana*, where no SP-positive cell bodies were observed (Inagaki et al. 1981). It has, however, to be stressed that in the green frog, *Rana esculenta*, SP-positive cell bodies are seen in the striatum (Taverna and Fasolo, unpublished observations).

Enkephalins. Enkephalin (ENK)-like immunopositivity is encountered in fibers and cell bodies of the striatum. It is mainly localized in cells located in the vicinity of SP-positive ones, and in some cases coexistence of the two substances is also observed (Taverna and Fasolo, unpublished observations). In Sauropsida, SP and ENK have been found in two separate neuronal populations (Reiner et al. 1984; Anderson and Reiner 1987; Reiner and Anderson 1987).

Somatostatin. Somatostatin (SRIF)-positive neuronal systems have already been described in the striatum of the crested newt (Fasolo et al. 1981). Stained cell bodies are more frequent in the caudal portions of the striatum proper and in the amygdaloid complex. Abundant nerve fibers are also observed and the *stria medullaris* appears heavily stained.

Neuropeptide Y. In a recent paper by Perroteau et al. (1988), neuropeptide Y (NPY)-like fiber systems in the striatum have been described (Fig. 4), whereas immunopositive cell bodies were reported mainly in the amygdaloid region.

Monoaminergic systems. Monoamine-containing cell bodies have rarely been found in the basal telencephalon. On the contrary, significant monoaminergic inputs have been described on the basis of both classical histofluorescence and immunohistochemical techniques. In particular, the striatum receives serotonin positive afferents from the raphe and midbrain (Fasolo et al. 1986). Moreover, abundant tyrosine hydroxylase immunopositive nerve fibers (possibly corresponding to the amine-containing fibers shown by Falck's method) innervate the striatal neuropil and periventricular gray (Franzoni et al. 1986; Fig. 5). Using an anti-dopamine antiserum, Franzoni et al. (unpublished observations) found that part of these aminergic systems appear to arise from the mesencephalic tegmentum and thus represent the terminal portion of a mesencephalo-striatal pathway. Afferents from the aminergic cells of the paraventricular organ, forming a prominent hypothalamo-striatal pathway in Urodela, were seen after histochemical as well as experimental studies (Dubé and Parent 1982). The amine-containing cells of the mesencephalic tegmentum are surrounded by a thick SP-positive axonal network (Fasolo et al., unpublished observations) and SP-positive fibers arising from the striatum give rise to a striato-tegmental pathway.

GABA. GABA-positive structures are detected by immunohistochemical methods throughout the telencephalon (Franzoni and

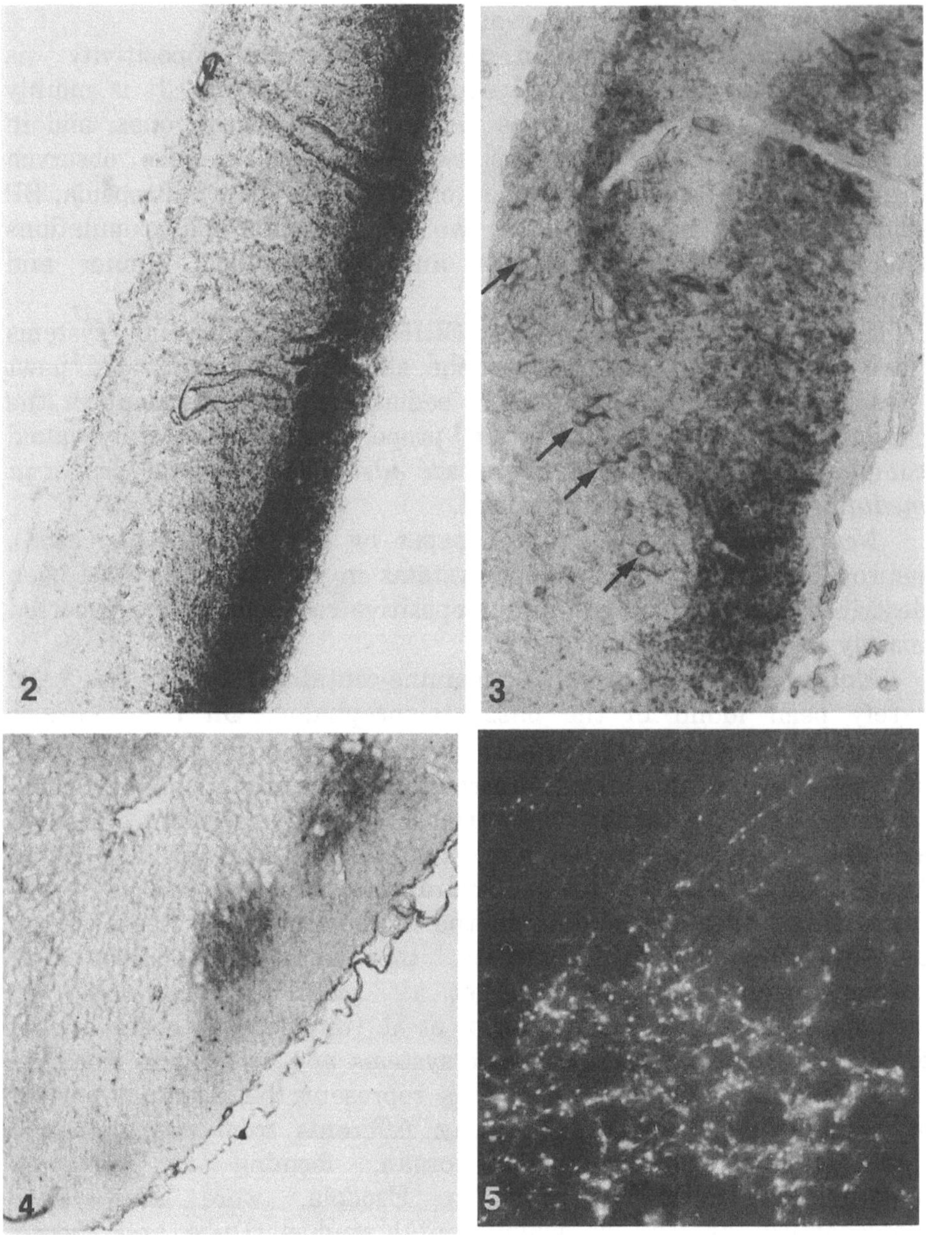

Fig. 2. Histochemical demonstrations of acetyl cholinesterase activity in the striatum of the newt. Transverse section. x 160

Fig. 3. Substance P immunopositive cell bodies (*arrows*) and fibers in the striatum of the crested newt. Avidin-biotin complex (ABC) method. Transverse section. x 120

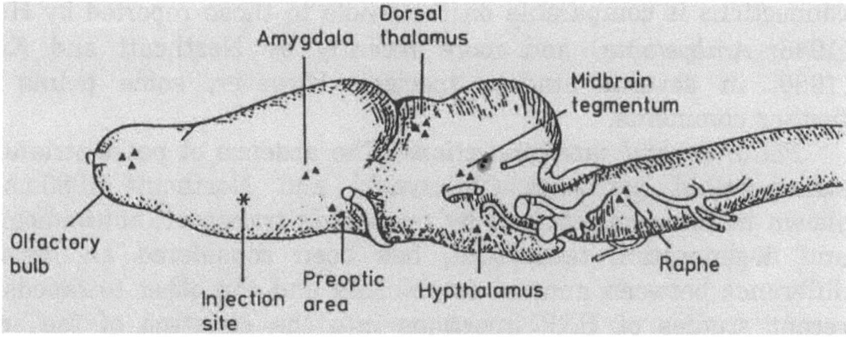

Fig. 6. Drawing of cell systems labeled by introduction of HRP into the striatum. Location of cell groups (▲) is shown as a projection onto the lateral aspect of the brain

Morino 1989). Some cell bodies are weakly immunolabeled in the striatal gray while the bulk of GABA-containing cells are located in the pallial areas. The medial and lateral forebrain bundles are intensely GABA-immunopositive.

Connections of the Striatum

In order to study the connections of the striatum in detail, HRP was injected or delivered by detergent soaked HRP-chips into the striatum or, alternatively, in some other areas, such as the pallium (lateral, dorsal and medial), habenulae and olfactory bulbs. To achieve a high sensitivity of peroxidase detection, an immunohistochemical staining against exogenous peroxidase was performed on sections from paraffin embedded brains.

In brief, the main afferents to the striatum arise from the olfactory bulbs and the so-called anterior olfactory nucleus, the septum, the pallium, the amygdaloid area, the dorsal thalamus, the preoptic area, the hypothalamus (mainly from its lateral regions), the tegmentum and the raphe (Fig. 6). This pattern of afferent

Fig. 4. Neuropeptide-Y immunopositive fibers and terminal fields in the striatum. ABC method. Transverse section. x 120

Fig. 5. Tyrosine hydroxylase immunopositive fibers innervating the newt striatum. Indirect immunofluorescence method. Transverse section. x 220

connections is comparable on the whole to those reported by Herrick (1948; *Ambystoma*) and more recently by Northcutt and Kicliter (1980; in several urodele species). However, some points need further comments.

Pallio-striatal interconnections. The absence of pallio-striatal and striato-pallial pathways (Wilczynski and Northcutt 1983a,b), as shown in *Rana catesbeiana* by retrograde transport, autoradiography and degeneration techniques, has been considered an important difference between anuran amphibians and the other tetrapods. Our recent studies of HRP injections into the striatum of the crested newt showed labeled cells in the dorsal, lateral and medial pallium. Such intriguing observations can be explained in several ways: (1) they might be an artifact due to HRP filling of passing fibers or, (2) the result of a higher sensitivity and better reliability of the immunohistochemical staining for HRP. Moreover, species differences may exist especially in such divergent groups as Urodela and Anura. However, applying the immunohistochemical method of tracing injected HRP to another anuran species, the green frog, we also found pallio-striatal connections.

Thalamo-striatal pathway. In the crested newt, the striatum receives an important projection from the dorsal thalamus. This result is in agreement with the observations on the thalamic connections in another closely related species of newt, *Triturus alpestris* (Wicht and Himstedt 1988). The origin of the pathway to the striatum is in a posterodorsal zone of the dorsal thalamus, whereas an anteroventral zone projects to the pallium.

Striato-hypothalamic relations. The striatum is connected to the preoptic and posterior hypothalamus. In turn, the posterior hypothalamus, and particularly the periventricular organ of the newt, exhibit important hypothalamo-striatal projections which may be involved in stereotyped motor behaviors of courtship (Dubé and Malacame, unpublished observations).

Conclusions

The use of multiple techniques to gain a deeper insight into the similarities between neural structures can be a powerful tool in comparative neurology, although some important theoretical issues on the homology concept require a more elaborate analysis (Fasolo and Malacarne 1988). In particular, the study of neurochemical and immunohistochemical markers can help to make more reasonable comparisons (Fasolo et al. 1988).

The neurochemical organization of the newt striatum shows several similarities with the immunohistochemical pattern exhibited in the lateral subpallial division of the African lungfish, *Protopterus annectens* (Reiner and Northcutt 1987). The latter structure, in turn, may be homologous to the basal ganglia of land vertebrates since it contains neuropeptide/neurotransmitter-specific neuronal populations (SP, ENK, and neurotensin-like cell populations; afferent aminergic and 5-HT fibers; an SP-positive bundle, possibly forming a striato-nigral pathway) that are characteristic for the striatal and pallidal portions of the basal ganglia in amniotes.

On the basis of their results, Reiner and Northcutt (1987) concluded that the evolution of the basal ganglia during the transition from fishes to amphibians must have been strikingly conservative.

The present results on the neurochemical organization and afferent connections suggest, on the whole, that the striatum of the urodele *Triturus cristatus* is basically organized like the striatal region in other tetrapods, and is, in many respects, less "aberrant" than that of ranid anurans.

This conclusion once more raises the problem of whether the urodele brain represents the *Typus* or the *Bauplan* of the vertebrate or at least of the tetrapod brain. A return to the concept of *Bauplan* might represent an intellectual breakthrough. The term *Bauplan* has been used with radically different meanings, i.e. as a way of conceptualizing the basic elements of the fundamental morphological pattern of the brain of vertebrates, or alternatively, in evolutionary terms, representing the brain of an actual ancestor.

At present, the available data cannot fully explain the mechanisms leading to the simple pattern of urodelan brain. Regression, neoteny, true primitivity bound to some developmental or evolutionary constraints might be considered for living species. Nonetheless, the study of the telencephalon in urodela using morphofunctional methods can supply some valuable clues to understanding the evolutionary trends of this region. In particular, analysis of the striatum can help to solve some intriguing problems, such as the origin of the dorsal ventricular ridge (Bissoli and Contestabile 1988; Contestabile and Bissoli, this volume).

References

Anderson KD, Reiner A (1987) Striatonigral projection neurons: a retrograde labeling study of the relative numbers that contain substance P or enkephalin. Soc Neurosci Abstr 13:3

Bissoli R, Contestabile A (1988) Evolution of neurotransmitter markers in vertebrate telencephalon. Comparative microchemical study in discrete brain regions of a frog and a turtle. Comp Biochem Physiol 89C:241-248

Bissoli R, Battistini S, Guarnieri T, Contestabile A (1988) Regional levels of neurotransmitter markers in the pigeon telencephalon: a comparison with possibly homologous areas of the rat telencephalon. J Neurochem 59:1731-1737

Dubè L, Parent A (1982) The organization of monoamine containing neurons in the brain of the salamander, *Necturus maculosus*. J Comp Neurol 211:21-30

Fasolo A, Franzoni MF, Gaudino G, Steinbusch HWM (1986) The organization of serotonin-immunoreactive neuronal system in the brain of the crested newt *Triturus cristatus carnifex Laur.* Cell Tissue Res 243:239-247

Fasolo A, Gaudino G, Mazzi V (1981) Somatostatin immunoreactive fibers and perikarya in the brain of the newt. 2. The extrahypothalamic pattern. Monit Zool Ital 15:229-238

Fasolo A, Malacarne G (1988) Comparing the structure of brains: implications for behavioral homologies. In: Jerison HJ and Jerison I (eds) Intelligence and evolutionary biology. Springer, Berlin, Heidelberg, New York, pp 119-141, NATO ASI Series 61(N7)

Fasolo A, Panzica GC, Viglietti Panzica C, Renda T, D'Este L (1988) Comparative chemical anatomy of the brain: concepts and methods. Basic Appl Histochem 32:15-30

Franzoni MF, Morino P (1989) The distribution of GABA-immunopositive neurons in the brain of the newt (*Triturus cristatus carnifex*) and the green frog (*Rana esculenta*). Cell Tissue Res 255:155-166

Franzoni MF, Thibault J, Fasolo A, Martinoli MG, Scaranari F, Calas A (1986) Organization of tyrosine hydroxylase immunopositive neurons in the brain of the crested newt *Triturus cristatus carnifex*. J Comp Neurol 251:121-134

Herrick CJ (1948) The brain of the tiger salamander. Univ Chicago Press, Chicago

Inagaki S, Senba E, Shiosaka S, Takagi H, Kawai Y, Takatsuki K, Sakanaka M, Matsuzaki T, Toyama M (1981) Regional distribution of substance-P like immunoreactivity in the frog brain and spinal cord: immunohistochemical analysis. J Comp Neurol 201:243-254

Northcutt RG, Kicliter E (1980) Organization of the amphibian telencephalon. In: Ebbesson SOE (ed) Comparative neurology of the telencephalon. Plenum, New York, London, pp 203-255

Parent A (1986) Comparative neurobiology of the basal ganglia. Wiley, New York

Perroteau I, Danger JM, Biffo S, Pelletier G, Vaudry H, Fasolo A (1988) Distribution and characterization of neuropeptide Y-like immunoreactivity in the brain of the crested newt. J Comp Neurol 275:309-325

Reiner A, Anderson KD (1987) The co-occurrence of substance P and GABA in striatal projection neurons of the basal ganglia. Soc Neurosci Abstr 13:3

Reiner A, Northcutt RG (1987) An immunohistochemical study of the telencephalon of the African lungfish *Protopterus annectens*. J Comp Neurol 256:463-481

Reiner A, Brauth SE, Karten HJ (1984) Evolution of the amniote basal ganglia. Trends Neurosci 7:320-325

Taban CH, Cathieni M (1983) Distribution of substance P-like immunoreactivity in the brain of the newt (*Triturus cristatus*). J Comp Neurol 216:453-470

Wicht H, Himstedt W (1988) Topologic and connectional analysis of the dorsal thalamus of *Triturus alpestris* (Amphibia, Urodela, Salamandridae) J Comp Neurol 267:545-561

Wilczynski W, Northcutt RG (1983a) Connections of the bullfrog striatum. Afferent organization. J Comp Neurol 214:321-332

Wilczynski W, Northcutt RG (1983b) Connections of the bullfrog striatum. Efferent projections. J Comp Neurol 214:333-343

Pallial Organization in *Xenopus laevis*: A Hodological and Immunohistochemical Study. The Development of a Structural Concept on Telencephalic Function

C.L. Veenman
Dept. of Anatomy and Neurobiology
University of Tennessee, Memphis
875 Monroe Avenue, Memphis, TN 38163, USA

What specific function of the telencephalon as an entity can be considered to be general for all vertebrates? In my opinion, Farr and Savage (1978) used the right paradigm to test the function of the telencephalon. They did the following experiment: Telencephalon ablated and nonablated goldfish were allowed to choose a leg in a T maze. Once they had developed a preference, they were given a conditioned stimulus from another situation, a flash related to an electric shock. The nonablated animals immediately changed their preference, the ablated animals did not. Apparently, animals need a telencephalon to make decisions in new situations. All other telencephalon related behaviors, arousal, responsiveness, exploration, discrimination, inhibition, disinhibition, reversal learning, fear, avoidance, caution, fearlessness, etc. (for review, see Ebbesson 1980) can be judged to be needed for this function, but subordinate. Having come to this conclusion, I proceeded to determine what the circuitry which gives the animal the ability to perform this function should look like.

I propose that the telencephalon can be subdivided into a **selection system** and a **sensory-motor system** (Fig. 1). Based on sensory information (<u>input</u>) entering the **sensory-motor system** a certain behavior is driven (<u>output</u>). As long as no changes in the environment occur, a certain sensoric stimulus will result in a certain behavior. When an essential change occurs, the **sensory-motor system** should be able to direct various other behaviors on the basis of the incoming information (<u>conversion motor options</u>). Also a gate via which the right behavior is directed should be opened. The basic process of this gate setting may be that within the **selection system** correlations (<u>free associations</u>) of present and

Experimental Brain Research Series 19

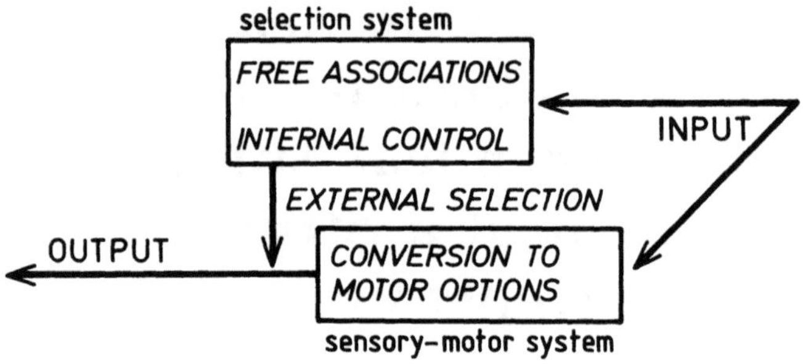

Fig. 1. A structural concept on telencephalic function. For explanation, see text.

past information (input) bring the **selection system** into a certain state of activity. Conflicting activity patterns will be weeded out (internal control) and finally from a stable activity pattern an outgoing train of activity will open a gate (external selection) for the output of the **sensory-motor system.**

The characteristics of the structural concept (Fig. 1) which can be tested are: Two parallel systems (a **selection system** and a **sensory-motor system**), which both receive multimodal sensorial information (input), where the **sensory-motor system** has direct access to motor regions (output) and the **selection system** can impinge on that outgoing information (external selection). The **selection system** should have a circuitry allowing for free associations and for internal control, i.e. on the one hand have a random circuitry and on the other hand show the possibility to block disorganized activity, whereas the **sensory-motor system** should have a logic circuitry for the conversion to motor options, i.e. it should somehow be clear that from a certain type of input only a restricted number of outputs can follow.

The model is based on research done on a bird, the goose (Veenman and Gottschaldt 1986), and a further development of the idea from Zeier (1971) that the efferent system of the telencephalon can be subdivided into a motor "effector" system and a limbic "gate setting" system. *Xenopus laevis*, an amphibian, was chosen as an example of a vertebrate with a relatively simple brain to see whether the basics of the model could be determined in a single species. The study presented here is treated extensively in a monograph (Veenman et al. 1989).

Observed *Xenopus laevis* spend most of their time sitting still under rocks under water. This behavior might seem too simple to imply a working telencephalon, but it is avoidance, and probably more strenuous than it seems at first sight; animals at ease remain

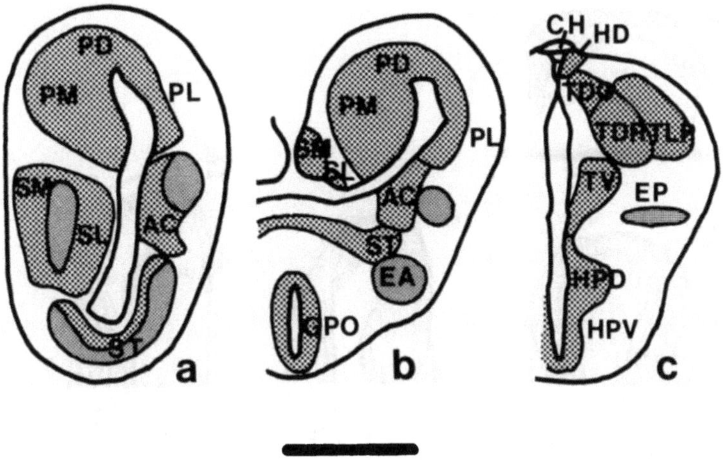

Fig. 2 a-c. Cross-sections through *Xenopus laevis* central telencephalon (**a**), caudal telencephalon (**b**), and diencephalon (**c**). *Bar*, 1 mm. For abbreviations, see text

at the surface, as do animals that are ill. In fact a very strong and relatively complex breathing reflex, involving a fast bout of swimming to the surface and back to the hiding place, must be suppressed. The general behavior of the animals is variable. Some animals come to the surface to be hand fed and in the course of weeks others will follow their example. When one animal is taken from the tank, none of the others will come to the surface from then on. So, the animals show individuality and the ability to learn, to reversal learning, and to respond adequately to new situations. *Xenopus* with its telencephalon removed is retarded in responding to events in its environment, although its locomotion patterns remain normal (K.E. Zittlau, personal communication). A male *Xenopus* with its forebrain removed is completely passive, but shows all the aspects of mating behavior when adequately stimulated (Hutchison and Poynton 1963/64). In a related species, *Rana catesbeiana*, telencephalic ablation results in the disappearance of the specific behavior elicited by unfamiliar situations (Stehouwer 1987). These data indicate that also in amphibians, including *Xenopus*, the telencephalon enables animals to make adequate decisions in new situations.

Xenopus laevis forebrain architecture is shown in Fig. 2. Most ventrally in the telencephalic hemisphere (Figs. 2a,b) is the striatum (ST). Dorsally are the medial (PM), dorsal (PD), and lateral (PL) areas of the pallium. Ventral to the medial pallium are the medial (SM) and the lateral (SL) septum. Laterally, in between

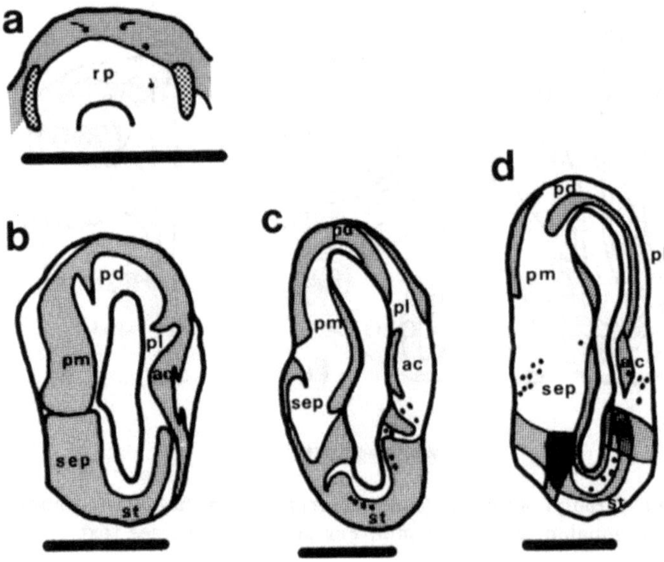

Fig. 3 a-d. Serotonergic (**a,b**), opioidergic (**c**), and substance P-ergic (**d**) cells (*dots*) and fibers (*shading*) in *Xenopus laevis* telencephalon. *Bars*, 1 mm. For abbreviations, see text

the pallium and the striatum, is the amygdaloid complex (AC), which on the basis of its position, connectivity and neurotransmitter content can be compared with the mammalian amygdala (Veenman et al. 1989). The telencephalic subdivisions form longitudinal zones (see also Fig. 5). Caudally (Fig. 2b) the striatum stretches over the midline, and the anterior entopeduncular nucleus (EA) replaces the striatal neuropil. The most rostral part of the hypothalamus is formed by the preoptic gray (GPO). Dorsally in the section of the diencephalon (Fig. 2c) are the dorsal habenula (HD) and the habenular commissure (CH). Different areas can be distinguished in the dorsal thalamus (TD): the caudal dorsal (TDC), rostral dorsal (TDR) and posterior lateral (TLP) thalamic nuclei. The ventral thalamus (TV) is a compact cell mass which is continuous with the hypothalamic gray. The hypothalamus (HT) can be subdivided into a dorsal (HPD) and a ventral part (HPV). Lateral to the gray of the ventral thalamus is the posterior entopeduncular nucleus (EP). The relative simplicity of *Xenopus* forebrain is reflected by the fact that the gray matter is generally periventricular while the white matter remains peripheral. Moreover, the gray matter shows few subdivisions.

Immunohistochemistry gives some more parameters to differentiate the telencephalic subdivisions (Fig. 3). Only in the

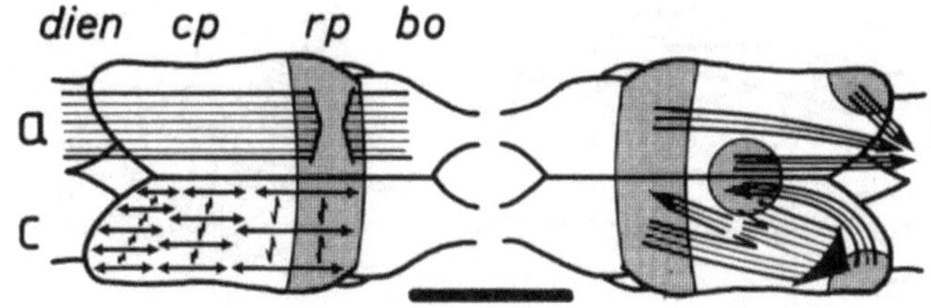

Fig. 4 a-d. Top views onto *Xenopus laevis* forebrain showing **a** afferents (*large arrows*) to the rostral pallium (*shaded*), **b** pallial sources (*shaded*) for efferents (*arrows*), **c** reciprocal connections (*double headed arrows*) within the pallium, **d** non-reciprocal connections (*arrows*) within the pallium. *Bar*, 2 mm. For abbreviations, see text

rostral pallium (rp) do serotonergic cells appear (Fig. 3a). A serotonergic fiber plexus appears throughout the telencephalic hemisphere (Fig. 3b). The medial pallium is permeated with this plexus, while it remains at the periphery of the dorsal and lateral pallium. Distinct longitudinal plexus reaching to the ventricle appear to demarcate the medial and lateral borders of the dorsal pallium (Figs. 3a,b). At the level of the amygdaloid complex the serotonergic plexus is displaced from the periphery by the lateral olfactory tract. The striatum and the septum (sep) are covered by one coherent field of serotonergic fibers and terminals. Opioidergic cells appear in the amygdaloid complex and the striatum (Fig. 3c). Substance P-ergic cells appear in the amygdaloid complex, the striatum, and the septum (Fig. 3d). This configuration indicates, when compared with the data reviewed by Nieuwenhuys (1985), that the amygdaloid complex can be compared with the mammalian amygdala. Dense fiber plexus for opioids and substance P (Figs. 3c,d) permeate the ventral part of the telencephalon, allowing differentiation of the septum, striatum and amygdaloid complex. Less dense plexus cover the pallium, allowing differentiation into medial, dorsal, and lateral parts. The rostral part of the pallium remains relatively free of substance P-ergic and opioidergic fibers (not shown).

HRP injections were used to elucidate the connectivity of the *Xenopus laevis* forebrain. Most injection sites were spread over the pallium. Further injections were placed in the olfactory bulb, the amygdaloid complex, the septum and the anterior entopeduncular nucleus. Some more injections were made in the thalamus, the hypothalamus, the optic tectum, and the ventral mesencephalon.

The rostral pallium receives afferents from the olfactory bulb (bo) and from the rostral dorsal thalamus (Fig. 4a). From the

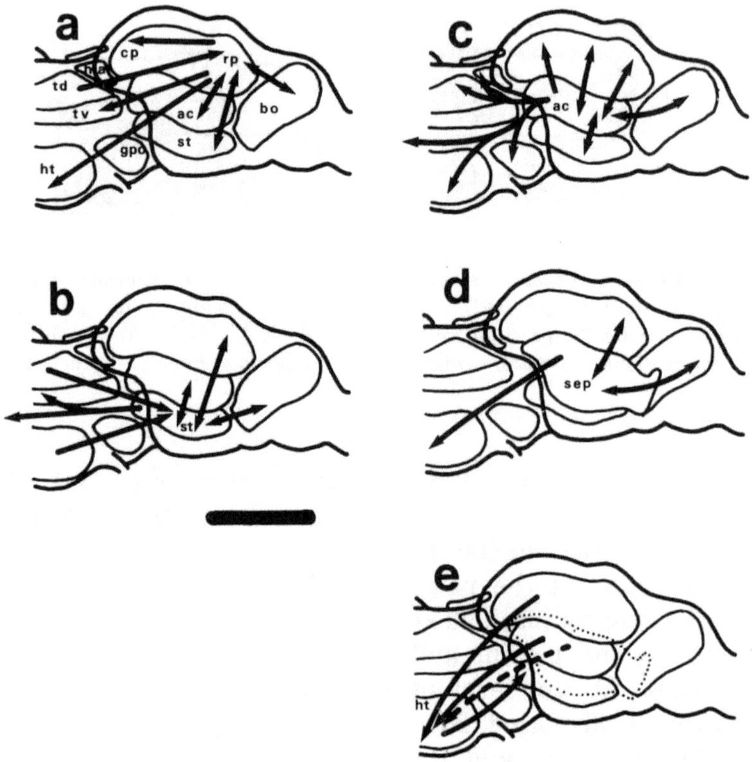

Fig. 5 a-e. Sagittal reconstructions of *Xenopus laevis* forebrain with the connections of **a** the rostral pallium, **b** the striatum complex, **c** the amygdaloid complex, **d** the septum, and **e** the hypothalamus. *Bar*, 2 mm. For abbreviations, see text

tectum to the thalamus in *Xenopus* different sensoric modalities are represented together (for review, see Veenman et al. 1989). So, the rostral pallium appears to be a multimodal input area for the caudal pallium (cp; see Fig. 4c). The pallium does not project beyond the diencephalon (dien; Fig. 4b). The entire pallium is covered with an intrinsic network of reciprocal connections which from rostral to caudal becomes more refined (Fig. 4c). The main input and output regions are interconnected with relatively long, nonreciprocal connections (Fig. 4d). Apart from the connections mentioned above the rostral pallium projects to the ventral thalamus and the hypothalamus and is reciprocally connected with the amygdaloid complex, the striatum and the septum (Figs. 5a,d). The olfactory bulbs are connected with all the telencephalic subdivisions 5a,b,c,d), the main targets being the rostral pallium, amygdaloid complex, and septum. The striatum complex (the

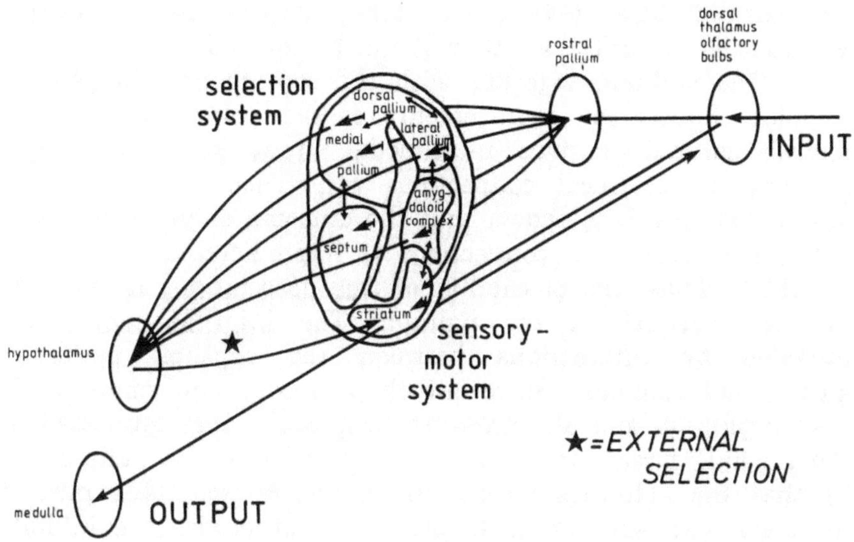

Fig. 6. The forebrain connectivity of *Xenopus laevis* related to the structural concept (Fig. 1). For explanation, see text

interconnected striatum and nucleus entopeduncularis anterior) receives afferents mainly from the accessory olfactory bulb, the caudal dorsal thalamus and the hypothalamus (Fig. 5b). The striatum is reciprocally connected with the rostral pallium and the amygdaloid complex. The anterior entopeduncular nucleus appears to be important relay for extratelencephalic connections of the striatum complex. The striatum is known to project as far as the spinal cord (ten Donkelaar et al. 1981). The amygdaloid complex (Fig. 5c) gets its main input from the olfactory system and sends efferents to the habenular nuclei, the medial dorsal thalamus, the medulla and to hypothalamic regions. Within the hemisphere it is connected with different pallial regions and the striatum. This connectivity is comparable with that of the mammalian amygdala (Nieuwenhuys et al. 1980). The connections of the septum were not extensively investigated (Fig. 5d). They show similarities to the connectivity of the amygdaloid complex (Fig. 5c). There is an olfactory input, output to the hypothalamus and connections with the pallium. The hypothalamus receives afferents from all pallial subdivisions, the septum and the amygdaloid complex and sends efferents especially to the nucleus entopeduncularis anterior of the striatum complex (Fig. 5e) and to the medulla.

The hodological data (summarized in Fig. 6) are in concordance with those obtained in *Xenopus* by Zittlau (personal communication) and parallel those obtained in other amphibians (for review, see Veenman et al. 1989). Further, it can be concluded:

1. The pallium, together with the septum and the amygdaloid complex, fulfills the requirements of the **selection system**. The rostral pallium receives multimodal input which by way of a network of reciprocal connections can reach all of the pallium. Mathematically it is shown that the outcome of processing within a network of reciprocal connections is unpredictable (Johannesma et al. 1986). Thus, the possibility of free associations is given by the intrinsic network of the pallium. The internal control can be provided by interactions between the septum, pallium, and amygdaloid complex. Since all three receive information via the rostral pallium and all three are reciprocally interconnected, as are the medial, dorsal, and lateral subdivisions of the pallium, it may be that the activities within the different structures must be in harmony with each other to persist, or otherwise are inhibited by mutual interactions.

2. The striatum complex fulfills the requirements of the **sensory-motor system**. It gets a clear multimodal input and has a homogeneous intrinsic connectivity and a direct output to motor regions in lower brain parts, which is necessary for conversion to motor options. The organization of its neuropil, where incoming axons, protruding dendrites and outgoing axons mix in a straightforward hardware configuration, like described for the tiger salamander by Herrick (1948), indicates that processing in this system is fairly rigid. This means that information entering this system will only result in a restricted number of outcome possibilities.

3. The hypothalamus takes a crucial position in the pathway for external selection from the **selection system** to the **sensory-motor system**. Possibly, different activity trains from the septum, pallium, and amygdaloid complex are weighed against each other within the hypothalamus; the resultant activity can, by way of the projection to the anterior entopeduncular nucleus and the striatum itself, determine what output will leave the striatum complex. That such processes indeed take place in hypothalamic regions in response to telencephalic stimulation was shown in the sunfish by Hallowitz et al. (1971).

So, the structural concept on telencephalic function can be recognized in the morphology of *Xenopus laevis* forebrain. More physiological support for the concept comes from work of Finkenstaedt and Ewert (1987) who showed that in habituated toads the ventral medial pallium, preoptic areas, and the dorsal

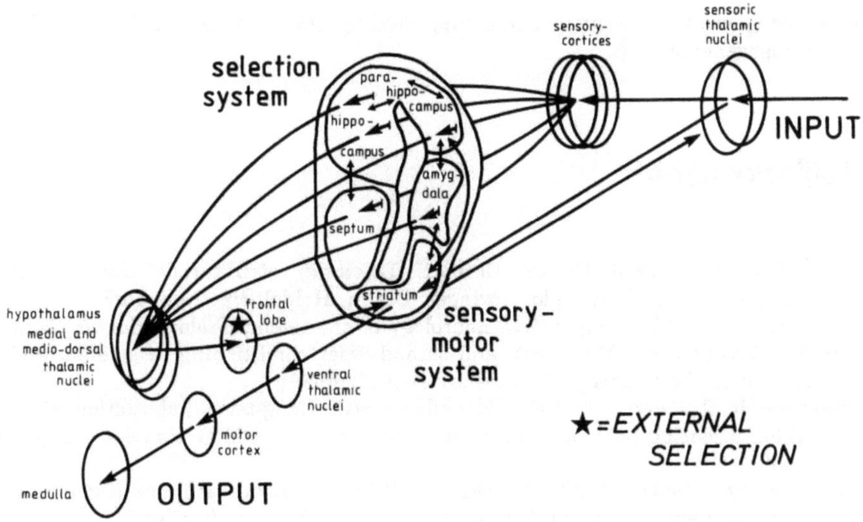

Fig. 7. A simplified diagram of forebrain connectivity of *Homo sapiens* as an example for mammals. Compare with Figs. 1 and 6. For reference, see text

hypothalamus (parts of the conceptual **selection system**) are active, whereas the striatum (the core of the conceptual **sensory-motor system**) and the lower brain parts (midbrain tectum, tegmental reticular formation and the ventral cerebellum) are inactive. The concept predicts that when the **selection system** blocks the **sensory-motor system**, then the first must be active while the second plus the lower brain parts are inactive.

To see whether the model is generally applicable, it was further tested with data collected on other vertebrates. As an example data from mammals are taken. A larger survey including fishes, amphibians, reptiles, and birds is given by Veenman et al. (1989). In Fig. 7, where the same layout is used as in Fig. 6, the data of mammals are assembled to see whether they can be compared with the pattern observed for *Xenopus laevis*. The mammalian data come mainly from two works, namely, "Cortex Cerebri" (Creutzfeldt 1983) and "Das Zentralnervensystem des Menschen" (Nieuwenhuys et al. 1980). More recent, additional data are reviewed by Veenman et al. (1989).

General conclusion: Since the data of mammals (Fig. 7) can be drawn into a diagram developed on the basis of data collected on *Xenopus laevis* (Fig. 6), and thereby fit into the developed model (Fig. 1), the model has a degree of general applicability.

Acknowledgements. This project was supported by grant Ve 101/1-1 from the Deutsche Forschungsgemeinschaft.

References

Creutzfeldt OD (1983) Cortex Cerebri. Leistung, strukturelle und funktionelle Organisation der Hirnrinde. Springer, Berlin, Heidelberg, New York

Ebbesson SOE (1980) Comparative neurology of the telencephalon. Plenum, New York

Farr EJ, Savage GE (1978) First- and second-order conditioning in goldfish and their relation to the telencephalon. Behav Biol 22:50-59

Finkenstaedt T, Ewert JP (1987) Stimulus-specific long-term habituation of visually guided orienting behavior toward prey in toads: a 14-2DG study. J Comp Physiol A 163:1-11

Hallowitz RA, Woodward DJ, Demski LS (1971) Forebrain activation of single units in preoptic area of the sunfish. Comp Biochem Physiol 40A:733-741

Herrick CJ (1948) The brain of the tiger salamander *Ambystoma tigrinum*. University of Chicago Press, Chicago

Hutchison JB, Poynton JC (1963/1964) A neurological study of the clasp reflex in *Xenopus laevis* (Daudin). Behaviour 22:41-63

Johannesma P, Aertsen A, van den Boogaard H, Eggermont J, Epping W (1986) From synchrony to harmony: ideas on the function of neural assemblies and on the interpretation of neural synchrony. In: Palm G, Aertse A (eds) Brain theory, Springer, Berlin, Heidelberg, New York, pp 25-47

Nieuwenhuys R (1985) Chemoarchitecture of the brain. Springer, Berlin, Heidelberg, New York

Nieuwenhuys R, Voogd J, de Boer-van Huizen (1980) Das Zentralnervensystem des Menschen. Springer, Berlin, Heidelberg, New York

Stehouwer DJ (1987) Effect of tectotomy and decerebration of spontaneous and elicited behavior of tadpoles and juvenile frogs. Behav Neurosci 101:378-384

ten Donkelaar HJ, de Boer-van Huizen R, Schouten FTM, Eggen SJH (1981) Cells of origin of descending pathways to the spinal cord in the clawed toad *(Xenopus laevis)*. Neuroscience 6:2297-2312

Veenman CL, Gottschaldt KM (1986) The nucleus basalis - neostriatum complex in the goose (*Anser anser* L.). Adv Anat Embryol Cell Biol 96

Veenman CL, Crzan D, Kern H, Rickman M, Wahle P, van Mier P (1989) The anatomical substrate for telencephalic function. Adv Anat Embryol Cell Biol 117

Zeier H (1971) Archistriatal lesions and response inhibition in the pigeon. Brain Res 31:327-339

Organization of the Dorsomedial Cortex in the Lizard *Podarcis hispanica*

F.J. Martinez-Guijarro, E. Desfilis and C. Lopez-Garcia
Catedra de Citologia e Histologia
Facultad de Ciencias Biologicas, Universidad de Valencia
E-46100 Burjasot, Spain

The cerebral cortex of reptiles comprises medial (MC), dorsal (DC) and lateral (LC) areas. The cerebral cortex of lizards and snakes includes an additional cortical area, the dorsomedial cortex (DMC). All cortical areas show a three-layered pattern with an intermediate cell-rich layer (CL) sandwiched between outer (OPL) and inner (IPL) plexiform layers that contain sparsely distributed neurons. This pattern is similar to that of the archicortex in mammals (Carpenter and Sutin 1983). The dorsomedial cortex is the main target of a well-developed Timm-positive fiber system (Molowny 1980; Lopez-Garcia et al. 1983). The non-squamate reptiles so far investigated (i.e. turtles of the genera *Testudo* and *Pseudemys*) lack both a Timm-positive fiber system and DMC (Molowny 1980).

In lizards, the DMC receives afferents from all cortical areas (Lohman and Mentink 1972; Lohman and Van Woerden-Verkley 1976; Ulinski 1976; Bruce and Butler 1984; Martinez-Garcia et al. 1986; Lopez-Garcia and Martinez-Guijarro 1988; Martinez-Garcia and Olucha 1988). Extracortical afferents reach DMC from the mammillary bodies, the raphe nucleus and the dorsolateral anterior nuclei of the thalamus (Lohman and Van Woerden-Verkley 1978; Martinez-Garcia and Olucha 1988).

Afferents to the DMC are stratified. The target fields of the MC projection correspond to the Timm-positive fields (Lopez-Garcia and Martinez-Guijarro 1988; Olucha et al. 1988), whereas projection fibers from the remaining cortical and extracortical areas terminate in the Timm-negative fields (Martinez-Garcia and Olucha 1988).

The neuronal population of the DMC has been studied in several squamate reptiles. A diversity of neuronal types has been described using Golgi impregnations (Ramón y Cajal 1917; Minelli 1966; Northcutt 1967; Ebbesson and Voneida 1969; Regidor et al. 1974; Lacey 1978; Wouterlood 1981; Guirado et al. 1984; Berbel 1988). The bipyramidal neurons of the cell layer are the most conspicuous

Experimental Brain Research Series 19

neuronal type in the DMC, appearing in all studied species. In the plexiform layers multipolar and stellate neurons are the most frequently described neuronal types.

In this study, the lamination and neuronal population of the DMC in the lizard *Podarcis hispanica* are described using Timm-staining, HRP injections, electron microscopy, and Golgi impregnations. The results suggest that the organization of the DMC in lizards is similar to that of the CA3 hippocampal region in mammals.

Material and Methods

A total of 142 specimens of *Podarcis hispanica* (Sauria, Lacertidae) of both sexes were used in this study. Lizards were caught wild in the vicinity of our laboratory in Burjasot, Valencia. In the laboratory, animals were maintained in terraria under controlled conditions of light and temperature and fed a diet of *Tenebrio* larvae. Lizards were kept in the laboratory for periods not exceeding 1 month before experimentation.

The morphology of neurons in DMC was studied in sections from 130 Golgi-impregnated brains, processed according to the procedure described by Berbel (1986).

Horseradish peroxidase (HRP) was used to study the laminar organization of DMC in eight specimens. Some 5-10 nl of 20%-30% HRP (Sigma type VI) in 10% Saponin (Merck) were pressure injected in the dorsomedial (four specimens) and in the medial (four specimens) cortices using glass micropipettes (10 μm external tip diameter). After 1 day of survival, animals were anesthetized with ether and perfused through the heart. Four specimens were perfused with 1.25% glutaraldehyde and 1% paraformaldehyde in 0.12 *M* phosphate buffer, pH 7.2. The remaining four specimens were perfused with the same fixative saturated with H_2S to undergo Timm HRP double labeling (Lopez-Garcia and Martinez-Guijarro 1988). After being removed from the skull, brains were immersed in fixative for 2-3 h at 4°C. Brains were then transversely sectioned at 150 μm on a tissue chopper. Slices were developed for histochemical detection of HRP using diaminobenzidine as chromogen (La Vail and La Vail 1974), then osmium postfixed, and finally embedded in araldite using flat molds.

Brains of four specimens were processed for zinc detection at the light microscope (Timm staining; Danscher and Zimmer 1978) and electron microscope levels (sulphide-osmium procedure; Lopez-Garcia et al. 1984).

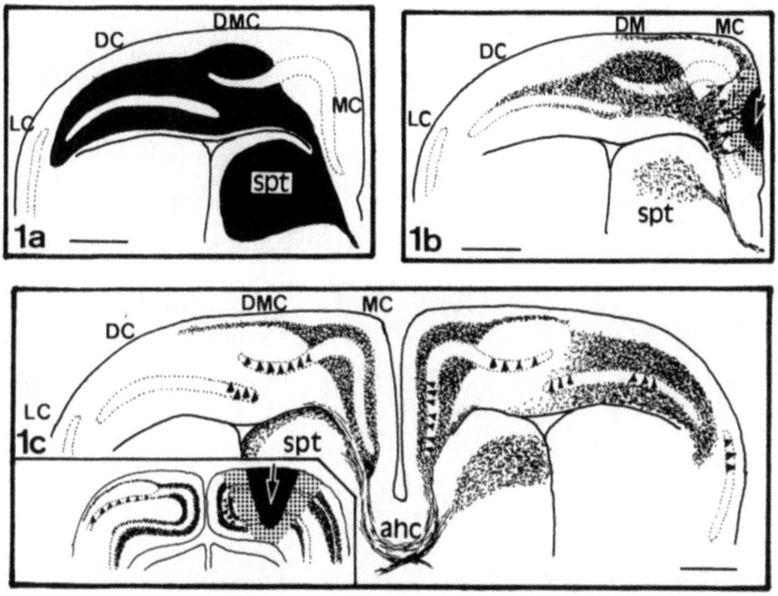

Fig. 1a-c. Drawings of transverse sections through the telencephalon of *Podarcis hispanica* at the level of the hippocampal commissure (*ahc*). **a** Timm-positive zones. **b** Anterograde labeling (*stippled zones*) after injection of HRP (*arrow*) into medial cortex (*MC*). **c** Anterograde labeling (*stippled zones*) after injection of HRP (*arrow in inset*) into the caudal part of dorsomedial cortex (*DMC*), also involving part of (*MC*). Labeling in dorsal cortex (*DC*) and septum (*spt*) corresponds to a projection from (*MC*). *LC*, lateral cortex; *dots*, extent of HRP diffusion; *triangles*, retrogradely labeled somata; *bars*, 250 μm

Results

Lamination

Timm Staining: Light and Electron Microscopy In lizards, Timm staining reveals a pattern of reactive cortical zones (Fig. 1a). Labeling is intense in the DMC except for the superficial third of the OPL, which is Timm-negative (Fig. 3a). The Timm-positive fields contain abundant Timm-positive boutons (TPB), easily identified in the electron microscope because their synaptic vesicles show an electrondense precipitate (Fig. 2a). TPB contain round clear vesicles, 40-45 nm in diameter and make asymmetric synaptic contacts, mainly with dendritic spines although they may also synapse on dendritic shafts or even neuronal somata.

Those boutons which are not labeled with Timm staining are called Timm-negative boutons (TNB). They make up a small

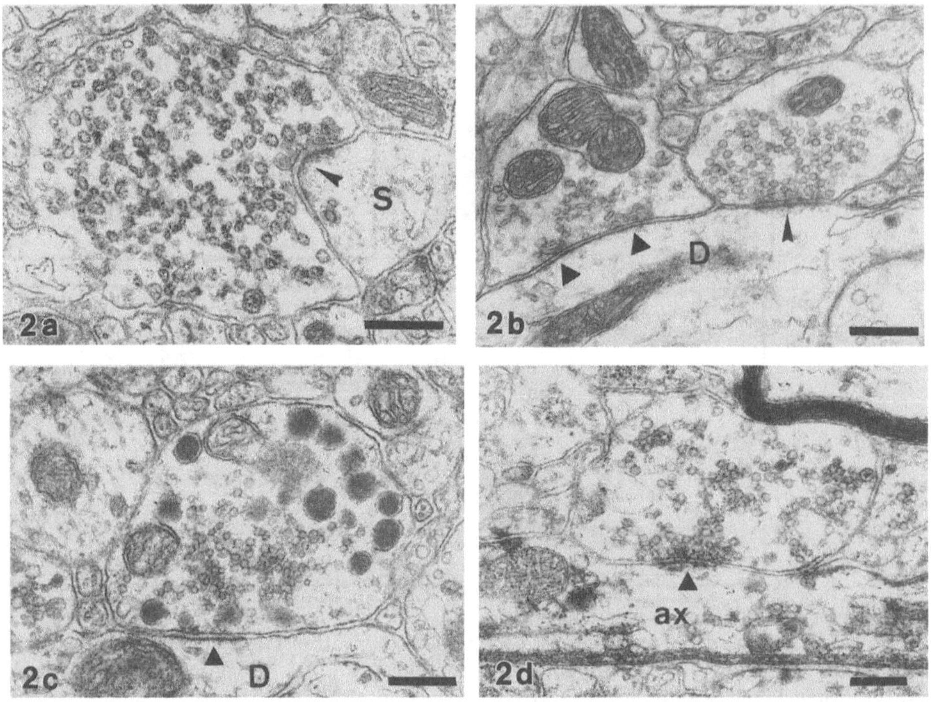

Fig. 2a-d. Types of synaptic boutons in the dorsomedial cortex. **a** Timm-positive bouton making an asymmetric contact (*arrowhead*) with a dendritic spine (*S*). **b** Dendrite (*D*) receiving asymmetric contact (*arrowhead*) from Timm-negative bouton with round, clear vesicles, and symmetric contact (*triangles*) from Timm-negative bouton with flattened vesicles. **c** Timm-negative bouton with small, round, clear vesicles and big dense-core vesicles making a synaptic contact (*triangle*) with a dendritic shaft (*D*). **d** Timm-negative bouton making symmetric axo-axonal contact (*triangle*). *ax*, axonal segment; *bars*, 0.25 μm

proportion of the synaptic boutons found in the Timm-positive zones of the DMC.

In the superficial Timm-negative field, two types of TNB are frequently found: TNB with round clear vesicles and TNB with flattened vesicles (Fig. 2b). About 90% of TNB contain round clear vesicles (20-30 nm in diameter) and make asymmetric synaptic contacts, usually on dendritic spines. TNB with flattened vesicles make symmetric synaptic contacts on dendritic shafts, and they represent about 10% of the boutons in the superficial Timm negative zone. These boutons are also found as boutons "en passant" making symmetric synaptic contacts on neuronal somata of the CL and on the thick proximal dendritic segments.

Two other types of TNB can be detected in the DMC: 1. TNB with small clear vesicles, 25 nm in diameter, and abundant dense-

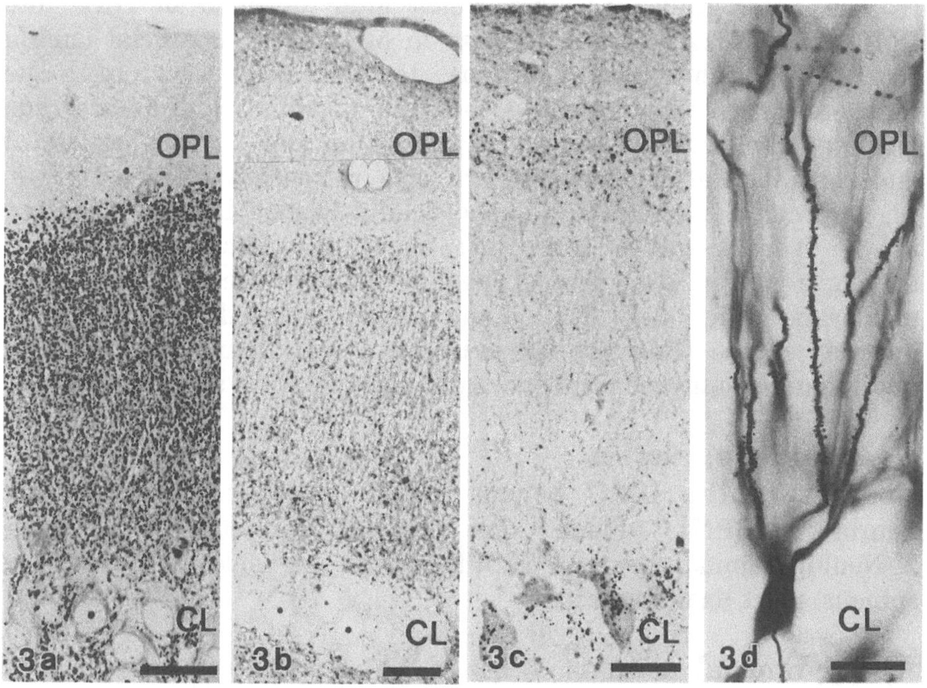

Fig. 3a-d. Lamination of dorsomedial cortex. **a** Timm-stained semithin section. The Timm-reactive zone occupies the inner two thirds of the outer plexiform layer (*OPL*). **b** Semithin section showing anterograde labeling after injection of HRP into medial cortex (see Fig. 1b). The projection field matches the Timm pattern. **c** Anterograde labeling restricted to the superficial third of the outer plexiform layer (*OPL*) following injection of HRP in the contralateral dorsomedial cortex (see Fig. 1c). Retrogradely labeled somata of bipyramidal neurons are seen in the cell layer (*CL*). **d** Golgi impregnation of bipyramidal neuron. The apical dendrites cross the termination zone of the projection from the medial and dorsomedial cortices. *Bars*, 20 μm

core vesicles with diameters of 80-120 nm (Fig. 2c); they are rare and make symmetric contacts on dendritic shafts in the plexiform layers. 2. TNB which make axo-axonal symmetric contacts on the initial axonal segments of neurons in the CL (bipyramidal neurons; Fig. 2d).

HRP Injections: Light and Electron Microscopy Injection of HRP into the medial (interhemispheric) part of MC results in anterograde labeling of the juxtasomatic laminae of DMC (Fig. 1b). This labeling matches the Timm staining pattern observed in this cortical area (Figs. 3a, 3b). In double-labeling Timm-HRP experiments, the HRP-labeled boutons are Timm-positive.

The anterograde projection fields after injection of HRP into DMC are two juxtasomatic laminae in MC and a superficial lamina in DMC which shortly extends to DC (Figs. 1c, 1d). The contralateral hemisphere shows the same distribution of anterograde labeling. The neuronal somata of the CL in DMC of the ipsilateral and contralateral hemispheres are also retrogradely labeled (Figs. 1c, 1d, 3c). In the DMC, the superficially labeled lamina is located in the Timm-negative zone, just above the Timm-positive field, leaving a narrow strip free of labeling beneath the superficial glial membrane (Fig. 3c). The anterogradely labeled boutons after injection in the DMC are Timm-negative with round clear vesicles and make asymmetric contacts mainly on dendritic spines.

Neuronal Population

Neurons in the DMC belong to two groups: 1. Bipyramidal neurons (BPN) which are spiny projection neurons. 2. Nonbipyramidal neurons (NBPN) which are smooth or sparsely spinous short axon neurons.

Bipyramidal neurons (Figs 3d, 4a, 4b) The cell body of BPN is large (23x16 μm) and is located in the CL. It gives off apical and basal dendrites which extend in two tufts through the OPL and IPL, respectively. The apical tuft is more developed than the basal tuft. Three to five thick and spine-free apical primary dendrites arise from the upper pole of the cell body. Secondary, tertiary and even quaternary dendrites have been observed. The latter are always found in the superficial third of the OPL. Secondary and higher order dendrites appear heavily covered with spines (Figs. 3d, 4a).

The number of spines on apical dendrites of BPN is very low in the initial dendritic segments, increases to reach a maximum (21.6 spines/10 μm of dendrite) in the intermediate dendritic segments (65 μm from the cell body), and decreases slightly in the distal tips (Fig. 4c).

Dendritic spines in the apical dendrites of BPN are of three types: stubby, thin and mushroom shaped (Fig. 4b). Thin and mushroom-shaped spines are located in the Timm-positive zone (intermediate and proximal dendritic segments) and are contacted by TPB. Most stubby spines are located in the Timm-negative zone, where they are contacted by TNB that form asymmetric synapses.

In the basal dendritic tuft, two or three primary dendrites branch near the cell body to give off secondary and tertiary dendrites. The basal dendrites are very spiny, and their tips usually

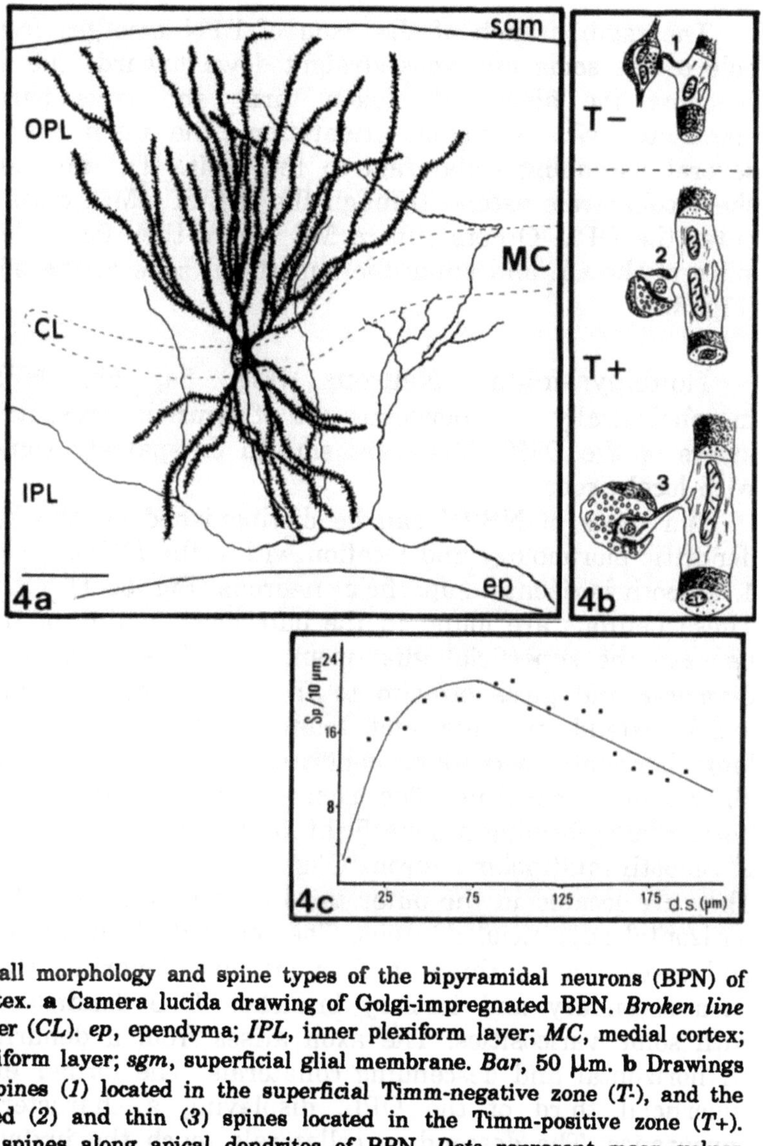

Fig. 4a-c. Overall morphology and spine types of the bipyramidal neurons (BPN) of dorsomedial cortex. **a** Camera lucida drawing of Golgi-impregnated BPN. *Broken line* delimits cell layer (*CL*). *ep*, ependyma; *IPL*, inner plexiform layer; *MC*, medial cortex; *OPL*, outer plexiform layer; *sgm*, superficial glial membrane. *Bar*, 50 µm. **b** Drawings of the stubby spines (*1*) located in the superficial Timm-negative zone (*T-*), and the mushroom-shaped (*2*) and thin (*3*) spines located in the Timm-positive zone (*T+*). **c** Frequency of spines along apical dendrites of BPN. *Dots* represent mean values obtained from 10 apical dendrites. *Solid line*, theoretical distribution obtained by adjusting the observed values to the Ruiz-Marcos and Valverde (1969) function. Chi-square test supports fit between the theoretical and the observed values ($P<0.05$). *SP/10 µm*, spines/10 µm of dendritic segment; *d.s.*, distance from the soma

do not reach the ependyma (Fig. 4a). BPN located in the lateral edge of the CL (which overlaps the CL of DC, forming the "superpositio medialis") generally show a distorted and poorly developed basal dendritic tuft.

The main branch of the axon of BPN emerges from the basal pole of the soma and goes straight down towards the alveus. Upon reaching the alveus the axon turns and runs parallel to the ependyma with a medial orientation. The main branch gives off several ascending collaterals to the DMC, DC and MC. Some of these collaterals ascend through the IPL of DMC, cross the CL and enter the OPL. Others run to MC where they form plexus on both sides of the CL and enter the superficial zone of the adjacent DMC (Fig. 4a).

Nonbipyramidal Neurons (Fig. 5a, 5b) **NBPN are a morphologically heterogeneous set of neurons present in all three layers of the DMC. They are smooth or sparsely spinous neurons with local axons.**

Ten types of NBPN can be distinguished on the basis of their dendritic morphology and location within the DMC.

1. Smooth horizontal superficial neurons (Fig. 5a-1)

These neurons are found in the outermost lamina of the OPL, just beneath the superficial glial membrane. The soma is 12-15 μm in diameter and gives off two to three horizontal primary dendrites which extend medially and laterally. Primary dendrites give off long, horizontal secondary dendrites. The dendrites are smooth or show small varicosities. The axon arises from the soma and extends horizontally, forming a superficial local plexus.

2. Smooth multipolar neurons (Fig. 5a-2)

They are located in the outer third of the OPL, beneath the smooth horizontal superficial neurons. The cell body is fusiform (16x11 μm) and gives rise to three to four primary dendrites that extend to form secondary and tertiary branches. The dendrites are smooth with some varicosities. The axon arises from a dendrite and gives off horizontal and descending collaterals. The former extend in the superficial third of the OPL, displaying a characteristic beaded appearance. The descending collaterals reach the juxtasomatic zone where they form a plexus parallel to the cell layer, showing numerous varicosities.

3. Sparsely spinous vertical multipolar neurons (Fig. 5a-3)

They are located in the OPL, near the cell layer. They have an ovoid, vertically polarized cell body of about 18x9 μm which gives rise to three or four primary dendrites. Near the soma, primary dendrites branch into secondary ones which take a vertical orientation, in such a way that the dendritic tree appears vertically arranged. In the OPL the dendrites almost reach the superficial glial membrane. Dendrites are sparsely spinous with varicose apical

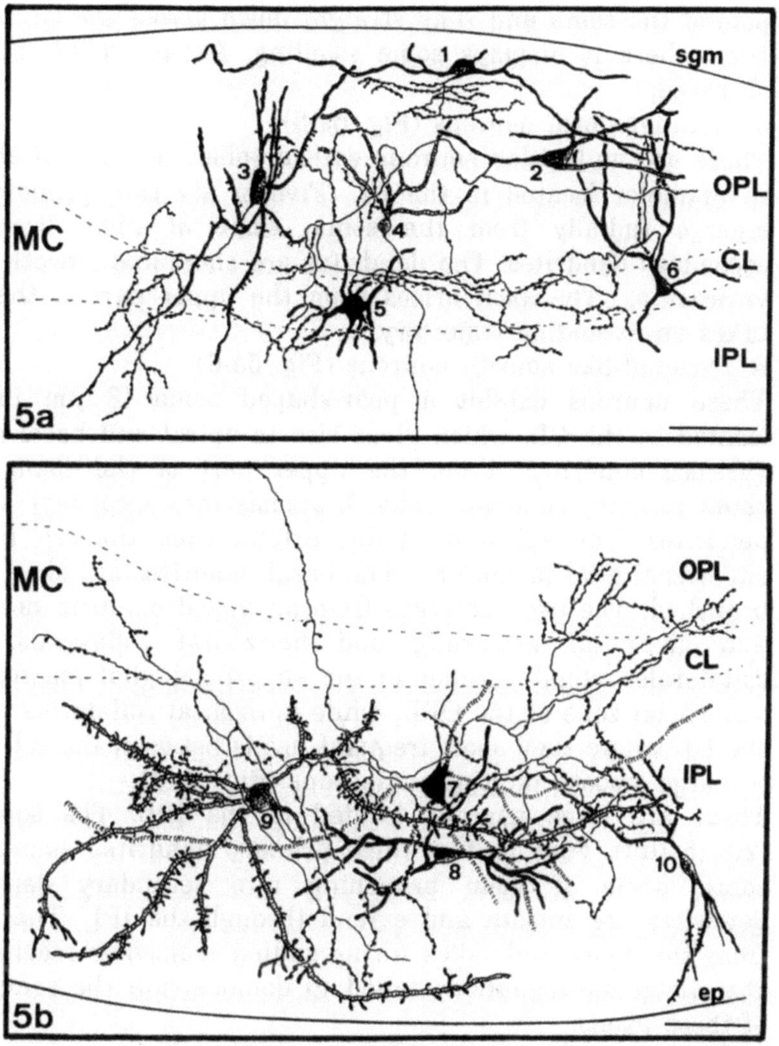

Fig. 5a,b. Non-pyramidal neurons in dorsomedial cortex. **a** Outer plexiform layer (*OPL*) and cell layer (*CL*). **b** Inner plexiform layer (*IPL*). For details, see text. *sgm*, superficial glial membrane; *ep*, ependyma; *bars*, 50 μm

tips. The axon emerges from a dendrite and descends to the CL where it ramifies, creating a plexus of varicose axonal segments.

4. Small smooth-tufted neurons (Fig. 5a-4)

They are neurons with a small soma (10 μm in diameter) located in the Timm-positive zone of the OPL. From the upper part of the soma emerges a single tuft of smooth or varicose dendrites which extends vertically through the OPL. The axon arises from the lower pole of the soma and runs straight down across the CL to enter the IPL, where it displays some swelling. Axonal collaterals were not observed.

5. Neurogliaform neurons (Fig. 5a-5)

These are multipolar neurons with a spherical soma of about 15 μm in diameter located in the CL. Five or six thin primary dendrites emerge radially from the soma, some of them branching into secondary dendrites. The dendrites are short and smooth with a few varicosities. The axon arises from the upper part of the soma and takes an ascending trajectory.

6. Pyramid-like smooth neurons (Fig. 5a-6)

These neurons exhibit a pear-shaped soma (8 μm in diameter) located in the CL, which gives rise to apical and basal smooth and varicose dendrites. From the upper part of the soma emerges a short primary dendrite which branches into secondary and tertiary dendrites. The apical dendrites extend from the cell layer to the superficial glial membrane. The basal dendrites are short and poorly branched. The axon emerges from an apical dendrite near the soma and gives off ascending and horizontal collaterals. Ascending collaterals extend parallel to the superficial glial membrane in the superficial zone of the OPL, while horizontal collaterals run through the CL where they show frequent beads between the cell bodies.

7. Large smooth multipolar neurons (Fig. 5b-7)

These are large neurons located in the IPL. The soma is ovoid (20x15 μm). Four or five thick primary dendrites emerge from the soma, some of them branching into secondary dendrites. The dendrites are smooth and extend through the IPL. The axon arises from the soma and takes an ascending trajectory reaching the CL. Our Golgi impregnations failed to demonstrate the complete extent of these axons.

8. Smooth bipolar horizontal neurons (Fig. 5b-8)

They are located in the IPL and have a spindle-shaped soma (11x8 μm) which gives rise to two or three horizontally arranged primary dendrites. These primary dendrites are thick and branch into secondary and short tertiary dendrites. The dendrites are generally smooth; some of them, however, bear a few sparsely distributed spines. The axon emerges from a primary dendrite and takes an ascending trajectory giving off non-varicose (i.e., "distribution") collaterals. Distribution collaterals give rise to varicose plexuses in the IPL, CL and in the juxtasomatic and superficial zones of the OPL.

9. Long-spined multipolar neurons (Fig. 5b-9)

These neurons are large and are located in the IPL. They show a medium-large soma (18-22 μm in diameter) from which four to five primary dendrites arise. At a variable distance from the soma, primary dendrites branch into secondary and tertiary dendrites. The dendritic tree extends into the IPL, largely with a rostro-caudal orientation. The dendrites are spiny, bearing particularly long spines distributed haphazardly along the dendrites. Branchlets can be seen arising from dendrites. The axon leaves the soma with a descending orientation, but soon turns to take an ascending course. The axon gives off distribution collaterals which run towards the CL of the MC and branch in the IPL of the DMC into beaded axonal segments.

10. Sparsely spinous bipolar vertical neurons (Fig. 5b-10).
These neurons are located in the IPL and are small in size. The cell body is thin and long (6x16 μm) and is vertically polarized. Two or three primary dendrites arise from the soma. Branching of the primary dendrites into secondary or higher order dendrites always takes place in the vicinity of the cell body. The dendritic tree shows a vertical distribution. The dendrites have sparsely distributed spines. The axon emerges from a basal dendrite with a descending orientation. Only the initial segment was impregnated in our material.

Discussion

In lizards, Timm staining labels the projection fields of the zinc-rich axonal system arising from the MC (Lopez-Garcia et al. 1983; Martinez-Guijarro et al. 1987; Molowny et al. 1987; Lopez-Garcia and Martinez-Guijarro 1988; Olucha et al. 1988).

The DMC receives two kinds of afferents: Timm-positive boutons (TPB) and Timm-negative boutons (TNB). These two kinds of afferents occupy different laminae and have different origins. TPB come from the MC and occupy the juxtasomatic fields. Their morphology is typical of excitatory boutons (round clear vesicles, asymmetric contacts). They have been compared to the mossy fiber boutons of the mammalian hippocampus (Molowny 1980, Lopez-Garcia et al. 1983; Martinez-Guijarro et al. 1984).

TNB are found in the superficial Timm-negative zone and in the CL. Since most of the intra- and extracortical afferents to the DMC converge on this superficial zone (Lohman and Mentink 1972; Lohman and Van Woerden-Verkley 1976; Martinez-Garcia et al. 1986; Martinez-Garcia and Olucha 1988), it is likely that TNB have diverse origins. Some TNB may be excitatory (round clear vesicles, asymmetric contacts; Uchizono 1965) and are usually found in the superficial zone. Others may be inhibitory (flattened vesicles, symmetric contacts) (Uchizono 1965) and are located both in the superficial Timm-negative zone and in the CL, where they form pericellular nets or make axo-axonal contacts. The distribution of the latter boutons parallels that of GABAergic boutons in this species (Schwerdtfeger and Lopez-Garcia 1986). From this, we suggest that TNB with flattened vesicles are probably GABAergic.

The distribution of presynaptic boutons in the DMC of lizards is similar to that in the CA3 hippocampal region of mammals (Gottlieb and Cowan 1972). In mammals, the Timm-positive mossy fibers terminate in the juxtasomatic stratum lucidum, while TNB occupy the stratum radiatum and the stratum lacunosum moleculare. The Timm-positive zone of the OPL of lizards can be compared to the stratum lucidum, while the superficial Timm-negative zone can be compared to the strata radiatum and lacunosum moleculare. In the lizard, as in mammals, zinc-rich afferents terminate in the juxtasomatic fields. The Timm-negative commissural and association afferents, on the other hand, reach the strata overlying the Timm-positive zone (Haug 1967, 1974; Laurberg 1979).

The neuronal population of the DMC of lizards is highly heterogeneous as seen in Golgi impregnations. Most neuronal types found in *Podarcis hispanica* have been previously reported for other species of squamate reptiles (Ramón y Cajal 1917; Minelli 1966; Northcutt 1967; Ebbesson and Voneida 1969; Regidor et al. 1974; Lacey 1978; Ulinski 1979; Wouterlood 1981; Guirado et al. 1984; Berbel 1988). However, this is the first time that all of them have been described in a single species. This suggests that the observed interspecific variability may be partly due the low number of brains studied in other works (between 15 and 42), since, at least in the present study, the number of neuronal types increased with the number of specimens examined.

The morphology of BPN is similar in *Podarcis hispanica* and in other lizards (Ramón y Cajal 1917; Minelli 1966; Northcutt 1967; Ebbesson and Voneida 1969; Regidor et al. 1974; Lacey 1978; Wouterlood 1981; Guirado et al. 1984, Berbel 1988), or even in snakes (Ulinski 1979). Moreover, the distribution of the number of spines along the apical dendrites of BPN of *Podarcis hispanica* is similar to that reported for other lizard species (Martinez-Guijarro et al. 1984).

The trajectories and arborization of the axons of BPN match the pattern of terminal labeling observed after injection of HRP in the

DMC. The labeling is bilateral and the neuronal somata of BPN in both the ipsi- and contralateral hemispheres are retrogradely labeled. Therefore, it is likely that BPN are responsible for both commissural and association connections projecting to the MC, DMC and DC. We may further hypothesize that BPN are probably excitatory since the labeled boutons seen after HRP injection in the DMC contain round clear vesicles and participate in asymmetric synaptic contacts.

BPN in the DMC receive Timm-positive afferents coming from the MC, which terminate on dendritic spines of the intermediate and proximal dendritic segments (Martinez-Guijarro et al. 1984), as well as Timm-negative afferents coming from other bipyramidal neurons located in the ipsilateral and contralateral hemispheres. The latter afferents terminate on distal dendritic spines located above the Timm-positive zone.

The distribution of afferents to BPN resembles that observed in the CA3 pyramidal neurons of mammals, which receive Timm-positive afferents from the fascia dentata (mossy fiber boutons) that end in the juxtasomatic stratum lucidum (Ramon y Cajal 1911; Blackstad and Kjaerheim 1961; Blackstad et al. 1970; Haug 1967, 1974). The CA3 pyramidal neurons of mammals also receive afferents from other pyramidal neurons of the CA3 region of both hemispheres which terminate in the stratum radiatum above the Timm-positive stratum lucidum (Laurberg 1979).

The BPN of lizards also project bilaterally to the juxtasomatic laminae of the MC, the area where PB originate. This projection is absent in the CA3 pyramidal neurons of mammals, which do not normally send collaterals to the fascia dentata. However, it has been recently reported using intracellular injection of HRP in slice cultures of rat hippocampus that pyramidal neurons of the CA3 send axonal collaterals to the fascia dentata (Frotscher and Gähwiler 1988). It is possible that the partially deafferentiated hippocampus acquires an organization more similar to that of the cortex of lizards. In any case, the projections of the BPN in lizards parallel those of some "hilar" neurons and CA3 pyramidal neurons of the hippocampus. Perhaps the BPN of the DMC retain both "hilar" and "CA3" roles in lizards.

The distribution of NBPN in the DMC matches that of GABAergic neurons described in this species, and their axonal arborizations show the same distribution as the GABAergic boutons (Schwerdtfeger and Lopez-Garcia 1986; Schwerdtfeger and Lorente 1988a,b). Therefore, most of the NBPN may be GABAergic neurons with local axons.

Some of the NPBN are probably primitive basket cells responsible for the axo-somatic synapses seen in the cell layer. Others may be axo-axonic cells which synapse on the initial axonal segment of BPN.

In summary, while BPN are projection neurons which make asymmetric synaptic contacts, probably excitatory, NPBN seem to be short axon interneurons, probably GABAergic and therefore inhibitory. These NBPN extend their dendrites across the projection fields of two presumably excitatory axonal systems coming from the MC and from BPN of the DMC, and can be contacted by boutons of these two fiber systems like the nonpyramidal GABAergic cells in the CA3 hippocampal region are innervated by mossy fiber boutons and commissural afferents (Frotscher 1985; Frotscher and Zimmer 1983; Frotscher et al. 1984). Therefore, NPBN in the DMC of lizards may participate in both feedforward and feedback inhibition phenomena.

In conclusion, the general laminar organization and the potential relationships of the neuronal elements with the afferent fibers in the DMC of lizards closely resemble those found in the hippocampal CA3 region of mammals.

Acknowledgements. We are indebted to Dr.E. Font for his comments on an earlier version of this manuscript. This work has been supported by the Spanish CAICYT and FISSS.

References

Berbel PJ (1986) Chromation at low temperatures improves impregnation of neurons in Golgi-aldehyde methods. J Neurosci Methods 17:255-259

Berbel PJ (1988) Cytology of medial and dorso-medial cerebral cortices in lizards: a Golgi study. In: Schwerdtfeger WK, Smeets W (eds) The forebrain of reptiles: current concepts of structure and function. Karger, Basel, pp 12-19

Blackstad TW, Kjaerheim A (1961) Special axo-dendritic synapses in the hippocampal cortex: electron and light microscopic studies in the layer of mossy fibers. J Comp Neurol 117:113-159

Blackstad TW, Brink K, Hem J, Jeune B (1970) Distribution of the hippocampal mossy fibers in the rat. An experimental study with silver impregnation methods. J Comp Neurol 138:433-450

Bruce LL, Butler AB (1984) Telencephalic connections in lizards. II. Projections to cortex. J Comp Neurol 229:585-604

Carpenter MB, Sutin J (1983) Human neuroanatomy. 8th edn. Williams and Wilkins. Baltimore, Maryland

Danscher G, Zimmer J (1978) An improved Timm-sulphide method for light and electron microscopic localization of heavy metals in biological tissues. Histochemistry 55:27-40

Ebbesson SOE, Voneida TJ (1969) The cytoarchitecture of the pallium in the tegu lizard *(Tupinambis nigropunctatus)*. Brain Behav Evol 2:431-466

Frotscher M (1985) Mossy fibers form synapses with identified pyramidal basket cells in the CA3 region of the guinea-pig hippocampus: a combined Golgi-electron microscope study. J Neurocytol 14:245-259

Frotscher M, Gähwiler BH (1988) Synaptic organization of intracellularly stained CA3 pyramidal neurons in slice cultures of rat hippocampus. Neuroscience 24:541-551

Frotscher M, Zimmer J (1983) Commissural fibers terminate on non-pyramidal neurons in the guinea pig hippocampus: a combined Golgi/EM degeneration study. Brain Res 265:289-293

Frotscher M, Léránth CS, Lübbers K, Oertel WH (1984) Commissural afferents innervate glutamate decarboxylase immunoreactive non-pyramidal neurons in the guinea pig hippocampus. Neurosci Lett 46:137-143

Gottlieb DI, Cowan WM (1972) On the distribution of axonal terminals containing spheroidal and flattened synaptic vesicles in the hippocampus and dentate gyrus of the rat and cat. Z. Zellforsch Mikrosk Anat 129:413-429

Guirado S, de la Calle A, Davila JC, Marin-Giron F (1984) Light microscopy of the medial wall of the cerebral cortex of the lizard *Psammodromus algirus*. J Morphol 181:319-331

Haug FMS (1967) Electron microscopical localization of the zinc in hippocampal mossy fiber synapses by a modified sulfide silver procedure. Histochemie 8:355-368

Haug FMS (1974) Light microscopical mapping of the hippocampal region, the pyriform cortex and the corticomedial amygdaloid nuclei of the rat with Timm's sulfide silver method. I. Area dentata, hippocampus and subiculum. Z Anat Entwicklungsgesch 145:1-27

Lacey DJ (1978) The organization of the hippocampus of the fence lizard: a light microscopic study. J Comp Neurol 182:247-264

Laurberg S (1979) Commissural and intrinsic connections of the rat hippocampus. J Comp Neurol 184:685-708

La Vail JH, La Vail MM (1974) The retrograde intraaxonal transport of horseradish peroxidase in chick visual system: a light and electron microscopic study. J Comp Neurol 157:303-358

Lohman AHM, Mentink GM (1972) Some cortical connections of the Tegu lizard *(Tupinambis teguixin)*. Brain Res 45:325-344

Lohman AHM, Van Woerden-Verkley I (1976) Further studies on the cortical connections of the Tegu lizard. Brain Res 103:9-28

Lohman AHM, Van Woerden-Verkley I (1978) Ascending connections to the forebrain in the Tegu lizard. J Comp Neurol 182:555-594

Lopez-Garcia C, Martinez-Guijarro FJ (1988) Neurons in the medial cortex give rise to Timm positive boutons in the cerebral cortex of lizards. Brain Res 463:205-217

Lopez-Garcia C, Soriano E, Molowny A, Garcia-Verdugo JM, Berbel PJ, Regidor J (1983) The Timm positive system of axonic terminals of the cerebral cortex of *Lacerta*. In: Grisolia S, Guerri C, Samson F, Norton S, Reinoso-Suarez F (eds) Ramon y Cajal's contribution to the neurosciences. Elsevier, Amsterdam, pp 137-148

Lopez-Garcia C, Molowny A, Perez-Clausell J, Martinez-Guijarro FJ (1984) A sulphide-osmium procedure for detection of metal containing synaptic boutons in the lizard cerebral cortex. J Neurosci Methods 11:211-220

Martinez-Garcia F, Amiguet M, Olucha F, Lopez-Garcia C (1986) Connections of the lateral cortex in the lizard *Podarcis hispanica*. Neurosci Lett 63:39-44

Martinez-Garcia F, Olucha FE (1988) Afferent projections to the Timm positive cortical areas of the telencephalon of lizards. In: Schwerdtfeger WK, Smeets W (eds) The forebrain of reptiles: current concepts of structure and function. Karger, Basel, pp 30-40

Martinez-Guijarro FJ, Berbel P, Molowny A, Lopez-Garcia C (1984) Apical dendritic spines and axonic terminals in the bipyramidal neurons of the dorsomedial cortex of lizards *(Lacerta)*. Anat Embryol (Berl) 170:321-326

Martinez-Guijarro FJ, Molowny A, Lopez-Garcia C (1987) Timm staining intensity is correlated with densities of Timm positive structures in the cerebral cortex of lizards. Histochemistry 86:315-319

Minelli G (1966) Architettura delle corteccie di alcuni rettili *(Lacerta muralis, Lacerta viridis, Testudo graeca, Crocodylus acutus)*. Arch Zool Ital 51:543-573

Molowny A (1980) Estudio de la corteza cerebral de *Lacerta* y otros reptiles con la técnica de Timm. Doctoral Thesis, Univiversity of La Laguna, Spain.

Molowny A, Martinez-Calatayud J, Juan MJ, Martinez-Guijarro FJ, Lopez-Garcia C (1987) Zinc accumulation in the telencephalon of lizards. Histochemistry 86:311-314

Northcutt RG (1967) Architectonic studies of the telencephalon of *Iguana iguana*. J Comp Neurol 130:109-148

Olucha FE, Martinez-Garcia F, Poch L, Schwerdtfeger WK, Lopez-Garcia C (1988) Projection from the medial cortex in the brain of lizards: correlation of anterograde and retrograde transport of horseradish peroxidase with Timm staining. J Comp Neurol 276:469-480

Ramón y Cajal P (1917) Nuevo estudio del encéfalo de los reptiles. Trab Lab Inv Biol Univ Madrid 15:83-99

Ramón y Cajal S (1911) Histologie du système nerveux de l'homme et des vertébrés. Maloine, Paris

Regidor J, Martín-Trujillo JM, Lopez-Garcia C, Marín F (1974) Estudio citoarquitectónico de la corteza cerebral de reptiles. II. Tipología dendrítica y distribución neuronal en areas corticales de *Lacerta galloti*. Trab Inst Cajal Invest Biol 66:1-32

Ruiz-Marcos A, Valverde F (1969) The temporal evolution of the distribution of dendritic spines in the visual cortex of normal and dark raised mice. Exp Brain Res 8:284-294

Schwerdtfeger WK, Lopez-Garcia C (1986) GABAergic neurons in the cerebral cortex of the brain of a lizard *(Podarcis hispanica)*. Neurosci Lett 68:117-121

Schwerdtfeger WK, Lorente MJ (1988a) GABA-immunoreactive neurons in the medial and dorsomedial cortices of the lizard telencephalon. In: Schwerdtfeger WK, Smeets W (eds) The forebrain of reptiles: current concepts of structure and function. Karger, Basel, pp 110-121

Schwerdtfeger WK, Lorente MJ (1988b) Laminar distribution and morphology of gamma-aminobutyric acid (Gaba-)immunoreactive neurons in the medial and dorsomedial areas of the cerebral cortex of the lizard *Podarcis hispanica*. J Comp Neurol 278:473-485

Uchizono K (1965) Inhibitory synapses in the stretch receptor neuron of the crayfish. Nature 214:215-225

Ulinski PS (1976) Intracortical connections in the snakes *Natrix sipedon* and *Thamnophis sirtalis*.J Morphol 150:463-484

Ulinski PS (1979) Intrinsic organization of the snake dorsomedial cortex: an electron microscopic and Golgi study. J Morphol 161:185-210

Wouterlood FG (1981) The structure of the mediodorsal cerebral cortex in the lizard *Agama agama*: A Golgi study. J Comp Neurol 196:443-458

Thalamo-Cortical Projections in the Lizard *Podarcis hispanica*

Fernando Martinez-Garcia and Maria-José Lorente
Universitat de Valencia, Facultat de Ciencies Biologiques, Departament de Biologia Animal, Unidad de Morfologia Microscopica, C/. Dr. Moliner, 50, E-46100 Burjassot, Spain

Studies on the thalamo-telencephalic projections in reptiles indicated that the cerebral cortex of turtles receives a visual projection from the lateral geniculate nucleus (Hall and Ebner 1970a,b; Ulinski 1988), whereas in lizards the only thalamic nucleus which projects directly to the cortex is the dorsolateral anterior nucleus (DLA) (*Tupinambis:* Lohman and Van Woerden-Verkley 1978; *Ophisaurus:* Belekhova and Kenigfest 1983; *Gekko:* Bruce and Butler 1984; *Podarcis and Gallotia:* Martinez-Garcia and Olucha 1988).

The functional significance of this thalamic projection in lizards is unclear. The DLA receives afferents from a variety of structures (Hoogland 1981, 1982) and thus probably relays to the cortex different kinds of information. Also our knowledge on the topographical organization of DLA projections is still poor. Therefore we have employed in this study the anterograde and retrograde transport of horseradish peroxidase (HRP) in order to determine the origin, course and termination mode of the ascending projection from the DLA to the cerebral cortex of the lizard, *Podarcis hispanica.*

Material and Methods

For this study 35 adult specimens of *Podarcis hispanica* were used. Of these, 28 received either injections of 15%-25% HRP and 10% saponin (Merck, Darmstadt, FRG.) or deposits of HRP-saponin crystals obtained by desiccating an aqueous solution of 15% HRP and 5% saponin) in different areas and subareas of the cerebral cortex. In these cases, the retrograde transport to the DLA was studied. In the remaining seven lizards HRP injections were placed in the thalamus in order to study anterograde transport to the cortex. After a survival time of 8-11 days, animals were either anesthetized and transcardially perfused with a fixative containing 1% paraformaldehyde and 1.25% glutaraldehyde in 0.1 *M* phosphate buffer, pH 7.2. Brains were removed and postfixed in the same fixative for 4 h at

Experimental Brain Research Series 19

4 C. After overnight storage in 20% phosphate buffered sucrose, 50 μm thick transverse sections were cut on a freezing microtome. For histochemical detection of HRP, we employed in matching series, respectively, a modified tetramethylbenzidine (TMB) method (Olucha et al. 1985) and the heavy metal enhanced diaminobenzidine (DAB) reaction (Adams 1981).

Results

The anterior thalamus of *Podarcis*

The DLA constitutes the largest nucleus of the anterior thalamus in *Podarcis*. In this species the DLA shows an oval profile in transverse sections (see Fig. 1A) and displays a more or less homogeneous population of neurons wherein, in contrast to other lizards (*Iguana:* Bruce and Butler 1984; *Gekko:* Smeets et al. 1986;), small-celled and large-celled parts cannot be distinguished. The DLA is laterally adjacent to the dorsomedial anterior nucleus of the thalamus (DMA) and ventrally adjacent to the habenula. At rostral levels, it is bounded laterally by the intercalate optic nucleus, and ventrally by the medial thalamic nucleus (Fig. 1A). At caudal levels it borders the dorsal optic nucleus (lateral to the DLA) and to the nucleus rotundus (ventral to the DLA; Fig. 1B).

Injections of HRP into the Cerebral Cortex

Whenever HRP was injected in the cerebral cortex, retrogradely labeled neurons were observed in the DLA (Fig. 1A). Labeling was bilateral after injections into the medial cortex and after rostral injections into any cortical area (Fig. 1A). On the other hand, an almost exclusively ipsilateral pattern of retrograde labeling was found in the DLA after restricted injections into the caudal dorsal cortex.

Labeled axons of DLA neurons coursed through the dorsal peduncle of the lateral forebrain bundle (LFB) at diencephalic levels. In the telencephalon, some of them entered the medial forebrain bundle (MFB) to end in the medial cortex. The remaining

→

Fig. 1A-C. A Retrogradely labeled somata in rostral DLA of *Podarcis* after ipsilateral HRP injection into rostral dorsal cortex. Fiber labeling of the stria medullaris is due to HRP diffusion to the lateral cortex. *Bar*, 200 μm. **B** Intense labeling of neurons in the caudal DLA in a specimen that received an ipsilateral deposit of HRP-saponin in the pallial thickening. Note that labeled dendrites do not enter the nucleus rotundus, while they are seen within the intercalate optic nucleus (*arrowheads*), where also

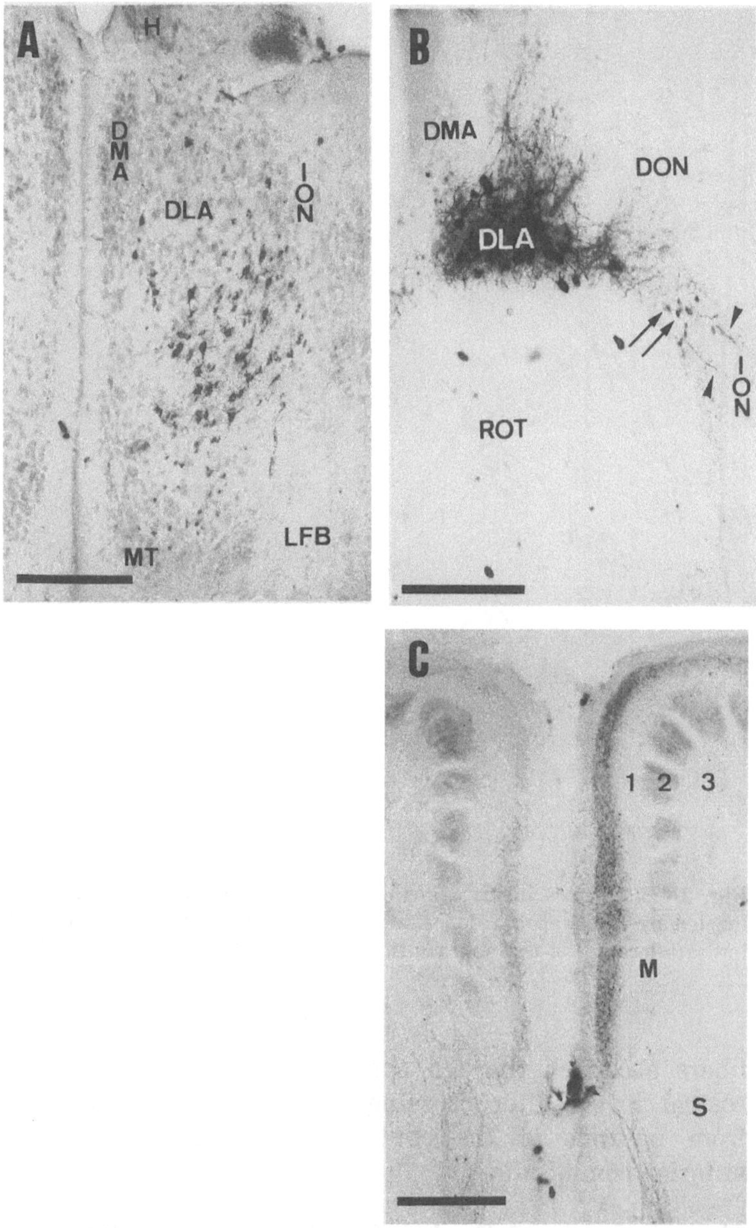

some labeled somata occur (*arrows*). *Bar*, 150 μm. **C** Terminal-like labeling in both medial cortices as a result of an HRP-saponin deposit restricted to the pallial thickening. *1, 2,* and *3*, layers of medial cortex. *Bar*, 150 μm. *DLA*, dorsolateral anterior nucleus; *DMA*, dorsomedial anterior nucleus; *DON*, dorsal optic nucleus; *LFB*, lateral forebrain bundle; *H*, habenula; *ION*, intercalate optic nucleus; *M*, medial cortex; *MT*, medial thalamic nucleus; *ROT*, nucleus rotundus; *S*, septum; *1*, outer plexiform layer; *2*, cell layer; *3*, inner plexiform layer

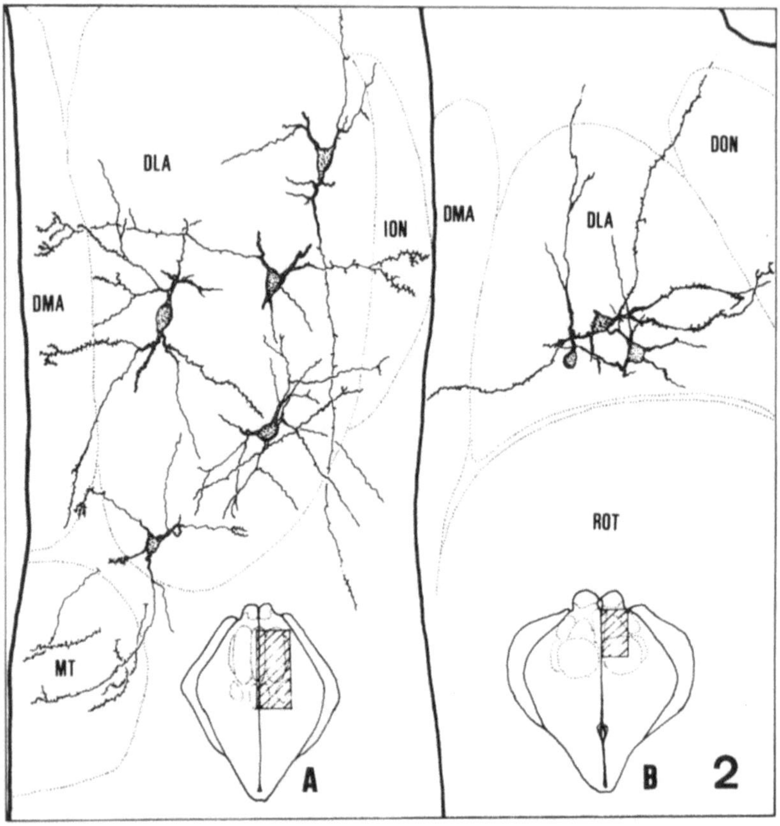

Fig. 2A,B. Camera lucida drawing of neurons in the rostral (**A**) and caudal (**B**) DLA labeled by HRP-saponin deposits in the telencephalon. Low power drawings (*bottom*) indicate location of the neurons in transverse sections. For abbreviations see Fig.1

fibers ascended through the LFB and entered the dorsal cortex at rostral levels after crossing the pallial thickening. Labeled fibers from neurons of the contralateral DLA decussated through the anterior commissure.

Morphology of Projection Neurons in DLA

When HRP-saponin crystals were deposited in the pallial thickening (at the entrance of the DLA fibers into the dorsal cortex) almost every ipsilateral DLA neuron showed intense labeling. Many of the cells were stained in a Golgi-like fashion (see Figs. 1B and 2). The labeled neurons showed a roughly stellate aspect with three to four main dendrites which ramified once or twice into sparsely spinous branches. The most noteworthy feature of these neurons was that

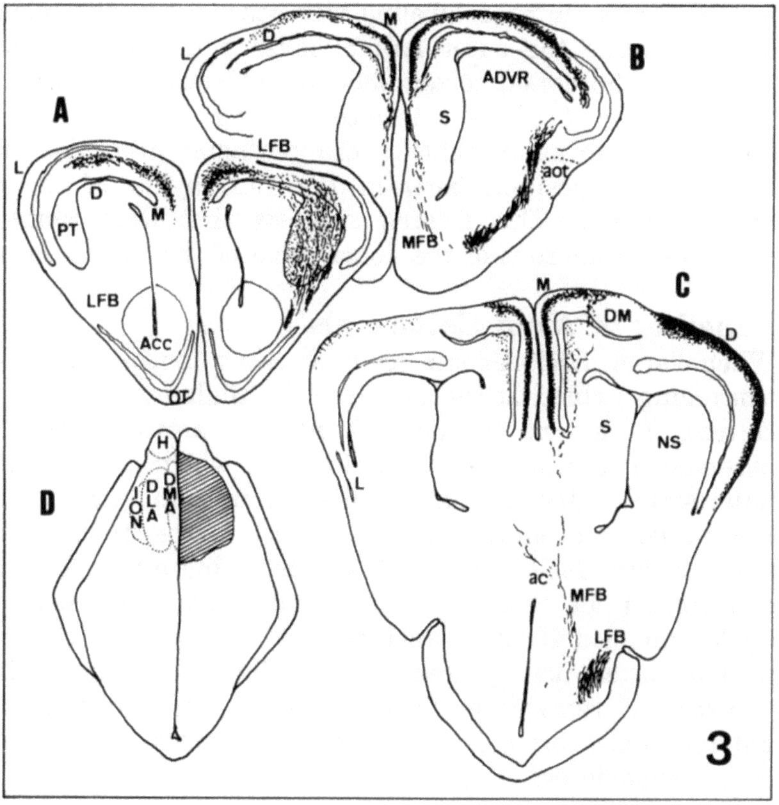

Fig. 3A-D. Camera lucida drawings of three transverse sections (**A**, **B** and **C**) through the telencephalon of a specimen that received an HRP injection into the anterior thalamus (**D**). Anterograde labeling of cortical structures is illustrated by *lines* (fibers) and *dots* (terminal-like labeling). *ac*, anterior commissure; *Acc*, nucleus accumbens; *ADVR*, anetrior-dorsal ventricular ridge; *aot*, accessory olfactory tract; *D*, dorsal cortex; *DLA*, dorsolateral anterior nucleus of the thalamus; *H*, habenula; *ION*, intercalate optic nucleus; *L*, lateral cortex; *LFB*, lateral forebrain bundle; *M*, medial cortex; *MFB*, medial forebrain bundle; *NS*, nucleus sphericus; *OT*, olfactory tubercle; *PT*, pallial thickening; *S*, septum

their dendrites exceeded the cytoarchitectonic boundaries of the DLA by extending highly spinous branches or tufts into adjacent thalamic nuclei (see Fig. 2), namely, the dorsomedial anterior, medial, dorsal optic and intercalate optic nuclei). The only adjacent nucleus that was not reached by dendrites from DLA neurons was nucleus rotundus (see Fig. 1B and 2B).

HRP Injections in the Thalamus

After restricted HRP injections into the anterior thalamus (Fig. 3), bilateral terminal-like labeling was observed in layers 1 and 3 of medial cortex. Intense labeling was observed in the middle third of layer 1, while the outer half of layer 3 displayed scarce labeling. Moreover, the dorsal cortex showed ipsilateral labeling in the outer half of layer 1 of its intermediate and lateral parts. At rostral levels, the terminal field of the thalamic projection occupied an intermediate position between the cell layers of the dorsal and lateral cortices. This finding suggests that DLA axons may form synapses on neurons of the dorsal zone of the rostral lateral cortex.

Labeling of Axon Collaterals

When large HRP-saponin crystals were deposited in the pallial thickening (which is the entrance for thalamic fibers to the dorsal cortex), a pattern of bilateral intense terminal-like labeling was observed in the medial cortex (Fig. 1C) which clearly resembled the anterograde labeling that resulted from thalamic injections. Since the pallial thickening does not project to the medial cortex (Bruce and Butler 1984; own unpublished findings), the most likely explanation for this labeling is that DLA neurons project to the ipsilateral medial and dorsal cortical areas and to the contralateral medial cortex using three axonic collaterals. Thus HRP could be transported retrogradely along one of these collaterals and then anterogradely through the other two towards the medial cortex in both hemispheres.

Discussion

Thalamo-Cortical Projections in Reptiles

The class *Reptilia* shows a great heterogeneity concerning the thalamo-cortical projections. In squamate reptiles and crocodiles, only one thalamic nucleus projects to the cortex, namely the DLA (Lohman and Van Woerden-Verkley 1978; Belekhova and Kenigfest 1983; Bruce and Butler 1984; Pritz and Stritzel 1987). In turtles two different groups of thalamic nuclei project directly to the cortex: the anterior thalamus including the dorsolateral anterior nucleus (Desan 1988) and the dorsomedial anterior nucleus (Balaban and Ulinski 1981) project to the medial cortex and to field D1 (medial part) of the dorsal cortex; the lateral geniculate nucleus projects exclusively to the lateral part (D2) of the dorsal cortex (Hall and Ebner 1970b; Desan 1988; Ulinski 1988). In view of both cytoarchitectonic and connectional data (Desan 1988) it can be assumed that the anterior thalamic nuclei of both reptilian groups are comparable. Hence the projection from the anterior thalamus to the archicortex may be a general feature of the reptilian brain.

Putative Role of DLA Neurons in the Processing of Sensory Information

In view of the dendritic organization that projection neurons of the DLA show, it is likely that one single DLA neuron may receive visual, somatosensory, auditory-vibrational and "visceral" information and forward it to telencephalic cortices in both hemispheres.

The dendritic branches of DLA neurons that enter the medial nucleus of the thalamus may receive auditory information from the torus semicircularis (Pritz 1974; Foster and Hall 1978). Those dendrites that invade the dorsal optic nucleus and the intercalate optic nucleus may be in synaptic contact with the retinal projection that terminates in these nuclei (Northcutt 1978). The retinal projection to the ipsilateral thalamus seems to reach the DLA as well (Cruce and Cruce 1978). Moreover, the DMA, which is also occupied by dendritic tufts belonging to DLA neurons, and DLA are the targets of an ascending projection from the spinal cord (Hoogland 1981). Finally, afferents of the DLA from hypothalamic nuclei (Hoogland 1982; Belekhova and Nemova 1987) may be considered to form part of the "visceral" sensory system.

These anatomical data are supported by electrophysiological findings in the lizard *Ophisaurus apodus* (Ivazov and Belekhova 1982; Belekhova and Ivazov 1983), according to which the DLA provides visual, audiovibrational and somatosensory input to the cerebral cortex.

Information about motor activity may reach the DLA indirectly from the striatum via nucleus entopeduncularis anterior (Hoogland 1977, 1982; Martinez-Garcia 1988; Russchen and Jonker 1988). The descending projections to the DLA from the septum (Hoogland 1982; Martinez-Garcia 1988) and from the dorsal cortex (Olucha et al. 1986; Martinez-Garcia and Olucha 1987; Hoogland and Vermeulen-Van der Zee 1988) are further sources of telencephalic influence on the activity of DLA neurons.

Despite the apparent lack of GABAergic neurons in the thalamic nuclei of reptiles that project to the telencephalon (Pritz and Stritzel 1988; own unpublished results) the data on fiber connectivity indicate that the thalamic DLA is not a simple "relay", but rather integrates sensory information with information from telencephalic structures about "motor" and, possibly, "limbic" events.

Comparisons

In spite of the differences found between the telencephalon of reptiles and mammals, there is now general agreement in comparing the medial and dorsomedial cortices of lizards (Hoogland and Vermeulen-Van der Zee 1988; Martinez-Garcia and Olucha 1988) and turtles (Desan 1988) with the mammalian hippocampal formation. One of the features that both structures share is the presence of afferents from the anterior thalamus. Although the main afferents of the mammalian anterior thalamus suggest that it may be considered as a relay in the visceral-sensory system (Herkenham 1978; Yanagihara et al. 1987), recently, a retino-thalamo-hippocampal projection has been found in the rat (Repérant et al. 1987). It is possible that also other sensory modalities could reach the mammalian hippocampus via the anterior thalamus, but no data exist on the neuronal substrate of such inputs.

There are, however, a few studies in other vertebrate classes on this topic. Wicht and Himstedt (1986, 1988) have analyzed the organization of the thalamo-telencephalic projections in a urodele, *Triturus alpestris*, and some similarities are obvious between their data and ours. The zona antero-ventralis of their dorsal thalamus displays a bilateral projection to the medial archipallium, which terminates in a similar way to the projection from the DLA to the medial cortex in lizards. In anurans, this projection reaches even the dorsal pallium (Neary 1984). Furthermore, the cells-of-origin of the thalamo-telencephalic pathway in Urodela extend dendritic branches into the adjacent retino-fugal neuropil of Bellonci (Wicht and Himstedt 1986, 1988).

Comprehensive studies concerning the neuronal typology, physiology and connectivity of the anterior thalamus in the different vertebrate groups are needed to understand the actual comparative and functional meaning of such a peculiar thalamo-telencephalic system.

Acknowledgements. The authors wish to thank Dr. Belekhova for her useful comments on the comparative physiology of the anterior thalamus. F.M.-G. has been awarded a predoctoral fellowship from the "Plan de Formacion de Personal Investigador".

References

Adams JC (1981) Heavy metal intensification of DAB-based reaction product. J Histochem Cytochem 29:775

Balaban CD, Ulinski PS (1981) Organization of thalamic afferents to anterior dorsal ventricular ridge in turtles. I. Projections of thalamic nuclei. J Comp Neurol 200:95-129

Belekhova MG, Ivazov NI (1983) Analysis of transmission of visual, somatic and audiovibrational sensory information to the hippocampal cortex of lizards. Neirofiziologiia 15:153-159

Belekhova MG, Kenigfest NB (1983) A study of the hippocampal mediodorsal cortex connections in lizards by means of horseradish peroxidase axonal transport. Neirofiziologiia 15:145-152

Belekhova MG, Nemova GV (1987) Study of connection of supposed limbic diencephalic nuclei in lizards using axonic HRP transport. Neirofiziologiia 19:110-120

Bruce LL, Butler AB (1984) Telencephalic connections in lizards. I. Projections to cortex. J Comp Neurol 229:585-601

Cruce JAF, Cruce WLR (1978) Analysis of the visual system in a lizard, *Tupinambis nigropunctatus*. In: Greenberg N, McLean PD (eds) Behavior and neurology of lizards. NIMH, Rockville, Maryland, pp 79-90

Desan PH (1988) Organization of the cerebral cortex in turtle. In: Schwerdtfeger WK, Smeets WJAJ (eds) The forebrain of reptiles. Karger, Basel, pp 1-11

Foster RE, Hall WC (1978) The organization of the central auditory pathways in a reptile, *Iguana iguana*. J Comp Neurol 178:783-832

Hall WC, Ebner FF (1970a) Parallels in the visual afferent projections of the thalamus in the hedgehog (*Paraechinus hypomelas*) and turtle (*Pseudemys scripta*). Brain Behav Evol 3:135-154

Hall WC, Ebner FF (1970b) Thalamotelencephalic projections in the turtle (*Pseudemys scripta*). J Comp Neurol 140:101-122

Herkenham M (1978) The connection of the nucleus reuniens thalami: evidence for a direct thalamo-hippocampal pathway in the rat. J Comp Neurol 177:589-610

Hoogland PV (1977) Efferent connections of the striatum in *Tupinambis nigropunctatus*. J Morphol 152:229-246

Hoogland PV (1981) Spinothalamic projections in a lizard, *Varanus exanthematicus*. An HRP study. J Comp Neurol 198:7-12

Hoogland PV (1982) Brainstem afferents to the thalamus in a lizard, *Varanus exanthematicus*. J Comp Neurol 210:152-162

Hoogland PV, Vermeulen-Van der Zee E (1988) Intrinsic and extrinsic connections of the cerebral cortex of lizards. In: Schwerdtfeger WK, Smeets WJAJ (eds) The forebrain of reptiles. Karger, Basel, pp 20-29

Ivazov NI, Belekhova MG (1982) Electrophysiological studies on afferent organization of the thalamus in the lizard *Ophisaurus apodus*. J Evol Biochem Physiol 18:76-85

Lohman AHM, Van Woerden-Verkley I (1978) Ascending connections to the forebrain in the tegu lizard. J Comp Neurol 182:555-594

Martinez-Garcia F (1988) Conexiones de las areas corticales y subcorticales del telencefalo de la lagartija comun *Podarcis hispanica*. Doctoral dissertation, University of Valencia, Burjasot

Martinez-Garcia F, Olucha F (1987) Thalamo-cortical reciprocal connection in the lizard *Podarcis hispanica*. Abstracts of International Brain Research Organization satellite symposium on recent advances in understanding the structure and function of the forebrain in reptiles. Frankfurt

Martinez-Garcia F, Olucha FE (1988) Afferent projections to the Timm-positive cortical areas of the telencephalon of lizards. In: Schwerdtfeger WK, Smeets WJAJ (eds) The forebrain of reptiles. Karger, Basel, pp 30-40

Neary TJ (1984) Anterior thalamic nucleus projections to dorsal pallium in ranid frogs. Neurosci Lett 51:213-218

Northcutt RG (1978) Forebrain and midbrain organization in lizards and its phylogenetic significance. In: Greenberg N, McLean PD (eds) Behavior and neurology of lizards. NIMH, Rockville, Maryland, pp 11-64

Olucha F, Martinez-Garcia F, Lopez-Garcia C (1985) A new stabilizing agent for the tetramethyl benzidine (TMB) reaction product in the histochemical detection of horseradish peroxidase. J Neurosci Methods 13:131-138

Olucha F, Poch L, Amiguet M, Martinez-Garcia F (1986) Differential connections of dorsal cortex subregions in lizards. Neurosci Lett [suppl] 26:S444

Pritz MB (1974) Ascending projections of a midbrain auditory area in a crocodile, *Caiman crocodilus*. J Comp Neurol 153:179-198

Pritz MB, Stritzel ME (1987) Percentage of relay and intrinsic neurons in a thalamic nucleus projecting to general cortex in reptiles, *Caiman crocodilus*. Brain Res 409:146-150

Pritz MB, Stritzel ME (1988) Thalamic nuclei that project to reptilian telencephalon lack GABA and GAD immunoreactive neurons and puncta. Brain Res 457:154-159

Repérant J, Weiner C, Pakhomova A, Desroches AM, Vesseklin N, Lemire M (1987) Mise en évidence d'une voie rétino-thalamo-hippocampique chez le rat. C R Acad Sci [III], 305:601-604

Russchen FT, Jonker AL (1988) Efferent connections of the striatum and the nucleus accumbens in the lizard *Gekko gekko*. J Comp Neurol 276:61-80

Smeets WJAJ, Hoogland PV, Lohman AHM (1986) A forebrain atlas of the lizard *Gekko gekko*. J Comp Neurol 254:1-19

Ulinski PS (1988) Functional architecture of turtle visual cortex. In: Schwerdtfeger WK, Smeets WJAJ (eds) The forebrain of reptiles. Karger, Basel, pp 151-161

Wicht H, Himstedt W (1986) Two thalamo-telencephalic pathways in a urodele, *Triturus alpestris*. Neurosci Lett 68:90-94

Wicht H, Himstedt W (1988) Topologic and connectional analysis of the dorsal thalamus of *Triturus alpestris* (Amphibia, Urodela, Salamandridae). J Comp Neurol 267:545-561

Yanagihara M, Niimi K, Ono K (1987) Thalamic projections to the hippocampal and entorhinal areas in the cat. J Comp Neurol 266:122-141

Postnatal Neurogenesis in the Brain of the Lizard *Podarcis hispanica*

C. Lopez-Garcia, A. Molowny, J.M. Garcia-Verdugo, F. Perez-Sanchez and F.J. Martinez-Guijarro
Catedra de Citologia e Histologia, Facultad de Ciencias Biologicas, Universidad de Valencia, E-46100 Burjasot, Spain

Postnatal neurogenesis and recruitment of newly formed neurons into functional circuits in the central nervous system of adult animals have been described for some vertebrate groups. Postnatal neurogenesis also plays an important role in the development of the retino-tectal system of fish and amphibians (Straznicki and Gaze 1971, 1972; Johns and Easter 1977; Stevenson and Yoon 1980; Johns 1982; Raymond and Easter 1983).

In songbirds, extensive postnatal neurogenesis occurs in sexually dimorphic forebrain nuclei (Nottebohm 1980). These nuclei are part of a pathway that has been implicated in the production of the species-typical song. Their growth during periods of active singing is hormonally mediated and may provide increased space for storage of the species-typical song, which is different every year (Nottebohm 1980; Goldman and Nottebohm 1983; Burd and Nottebohm 1985; Paton et al. 1985).

Postnatal neurogenesis in the mammalian central nervous system is restricted to short periods after the animal's birth. It has been observed in the olfactory bulbs (Hinds 1968a,b; Altman 1969; Kaplan and Hinds 1977; Rossellini-Austin and Altman 1979; Bayer 1983), hippocampal *fascia dentata* (Altman 1963; Altman and Das 1965; Angevine 1965; Schlessinger et al. 1975; Kaplan and Hinds 1977; Bayer 1980; Rakic and Nowakowsky 1981) and cerebellar cortex (Rakic 1971; Altman 1972a,b,c).

In reptiles, ^{3}H-thymidine autoradiography has been used to investigate neurogenesis in cortical forebrain structures during embryonic development (Goffinet et al. 1985) and also during postnatal life (Lopez-Garcia et al. 1988a,b). The latter studies confirmed that intense postnatal neurogenesis occurs in the medial cortex of the lizard, *Podarcis hispanica*. This result had been predicted by previous observations that the neuronal population of the medial cortex increases steadily during postnatal life in this species (Lopez-Garcia et al. 1984b). Furthermore, the presence of

Experimental Brain Research Series 19

immature cells attached to the vertical shafts of ependymocytes in the medial cortex of *Lacerta galloti* extends the range of species that exhibit postnatal neurogenesis and suggests a possible route for the migration of immature neurons (Garcia-Verdugo et al. 1986).

In the present study, we injected ^{3}H-thymidine into perinatal, juvenile and adult *Podarcis hispanica* to characterize brain structures that undergo postnatal neurogenesis in this species. The use of different survival times supplied information on the recruitment of postnatally generated neurons. Finally, an additional aim of this study was to provide evidence that the newly formed neurons are incorporated into functional cerebral circuits.

Material and Methods

Animals

Perinatal (1-30 days old, 24-25 mm head-cloaca length, HCL), juvenile (6-12 months old, 30-35 mm HCL) and adult (2-5 years old, 40-55 mm HCL) *Podarcis hispanica* (Sauria, Lacertidae) were used in this study. Lizards were wild caught in Burjasot (Valencia) during June (1985-87) and kept in laboratory for periods not exceeding 1 week before processing. In the laboratory, lizards were held in terraria and fed with *Tenebrio* larvae. Light and temperature simulated those in the animals' natural environment.

Treatment consisted of daily intraperitoneal injections of ^{3}H-thymidine (50 μCi/g of body weight) during three consecutive days. The animals were chilled in the refrigerator (2-5 min) to facilitate handling and injection of labeled DNA-precursor.

Autoradiography

Thirty lizards used in autoradiography experiments were allowed survival times of 1-47 days after the last ^{3}H-thymidine injection. The animals were then anesthetized with ether and perfused through the heart with buffered formalin. Their brains were removed from the braincases, immersed in Bouin's fixative and embedded in paraffin.

Cell counts were performed on 10 μm thick transverse serial sections of the brain. Microscope slides bearing the section were dipped in autoradiographic emulsion (Ilford K-5 nuclear emulsion diluted 1:1 in water), dried in the dark and stored at 4°C for 20-30 days. The slides were then immersed in Kodak D-19 developer for 5-8 min, rinsed in water, fixed in 20% sodium thiosulfate, counterstained with toluidine blue, dehydrated, cleared and mounted. Nuclei with a silver granule density exceeding three times that of the background were considered as marked nuclei.

Autoradiography-HRP Double Labeling

Nine animals were used in double labeling experiments. Two months after the last ^{3}H-thymidine injection, they received injections of the retrograde tracer horseradish peroxidase (HRP) into the dorsal cortex. Following survival times of 1-3 days, they were sacrificed and their brains processed for autoradiography and histochemical detection of HRP using diaminobenzidine as chromogen (for details see Lopez-Garcia and Martinez-Guijarro 1988).

Results

Radioactively labeled nuclei were consistently found in the olfactory bulbs, medial cortex, *nucleus sphericus* and cerebellar cortex in specimens of all ages (Fig. 1). The distribution of labeled nuclei in these four areas is shown in Figs. 2 and 3.

On the average, 24% of all labeled nuclei were found in the olfactory bulbs, 49% in the medial cortex, 4% in the *nucleus sphericus* and 12% in the cerebellum. Occasionally, a few labeled cells were observed in other areas (e.g. lateral cortex, dorsal ventricular ridge, striatum).

Olfactory Bulbs

In the main and accessory olfactory bulbs, most radioactively labeled nuclei were located in the ependyma and granular layers. A few nuclei of periglomerular cells in the glomerular layer and nuclei in the mitral layer were occasionally labeled.

Medial Cortex

In the medial cortex radioactively labeled cells were consistently observed in the ependyma and in the inner plexiform and granular layers. Rarely, a few labeled cells were found in the outer plexiform layer.

Nucleus sphericus

In *nucleus sphericus*, radioactively labeled cells were consistently observed in the ependymal, marginal and mural layers, but never in the central hilus.

Cerebellum

Radioactively labeled cells were found in all cerebellar layers.

Effect of Survival Time on the Location of Labeled Nuclei

Following short survival times after the last ^{3}H-thymidine injection, labeled cells were located in ependymal zones of the olfactory bulbs,

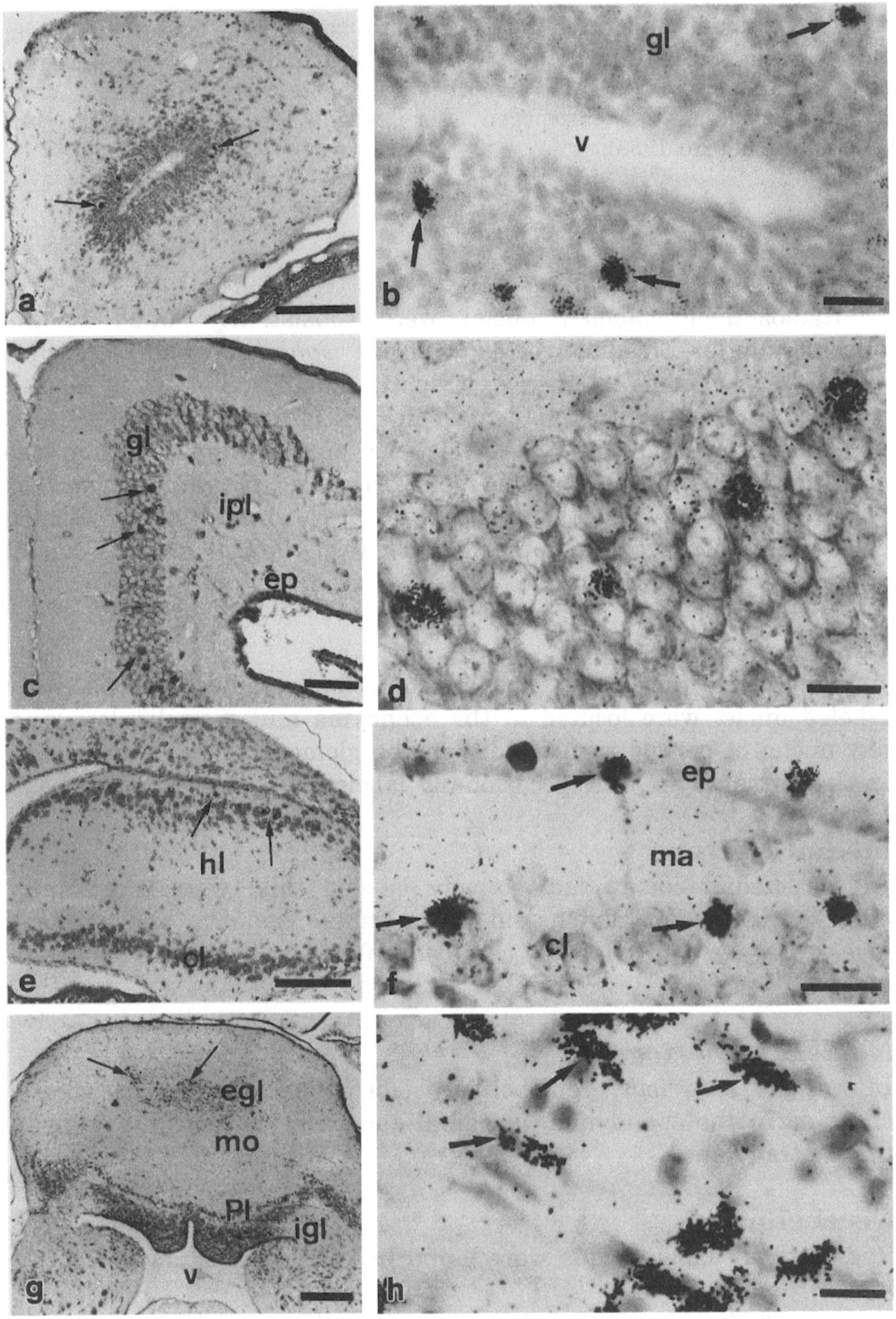
a
gl
v
b
gl
ipl
ep
c
d
hl
cl
e
ep
ma
cl
f
egl
mo
Pl
igl
v
g
h

medial cortex, *nucleus sphericus* and in the external granular layer of the cerebellum. These areas are generally regarded as germinative (Kirsche 1967).

Long survival times resulted in labeling of the granular or principal cell layers of telencephalic structures (i.e. the granular cell layer of the medial cortex, the mural cell layer of the *nucleus sphericus* and the outer part of the olfactory granular layers) and in the inner granular layer of the cerebellum (Fig. 2).

This shift in the frequency peak of labeled cells from the ependymal zones to the principal cell layers occurred between 5 and 18 days after the last ^{3}H-thymidine injection (see Fig. 2). During this "shift time", labeled cells were located in the intervening plexiform layers (i.e. the inner plexiform layer of the medial cortex, the marginal layer of the *nucleus sphericus* and the inner part of the granular layers in the olfactory bulbs). Frequency peaks for labeled cells in these plexiform layers occurred 5 and 18 days after injection in the medial cortex and in the *nucleus sphericus*, respectively. In the cerebellar cortex we detected two peaks 8 and 18 days after injection (Fig. 2).

In the main olfactory bulbs, the location of labeled nuclei within the granular layer depended on the survival time allowed after the last ^{3}H-thymidine injection. After 7 days, the majority of labeled nuclei were located in the inner part of the granular layer, whereas after 28 days the majority was detected in the outer part.

In the accessory bulb, labeled cells were preferentially located in juxtaependymal sites of the granular layer after short survival times. Long survival times resulted in labeled cells located farther away from the ependyma.

◄

Fig. 1a-h. a Transverse section through the main olfactory bulb of *Podarcis hispanica*. *Bar* = 100 µm. **b** Ependymal and granular layers of the main olfactory bulb. Autoradiographically labeled cells are located in the ependyma and in the granular layer (*gl*). *v* = ventricle; *Bar* = 25 µm. **c** Transverse section through the medial cortex. Labeled cell are located in the ependyma (*ep*), the inner plexiform layer (*ipl*) and the granular layer (*gl*). *Bar* = 50 µm. **d** Granular layer of the medial cortex. Some granule cells are heavily labeled. *Bar* = 10 µm. **e** Transverse section through nucleus sphericus. *hl* = hilus; *cl* = mural cell layer. *Bar* = 100 µm. **f** *Nucleus sphericus* (detail). Labeled cells can be observed in the ependyma (*ep*), the marginal layer (*ml*) and the mural cell layer (*cl*). *Bar* = 20 µm. **g** Transverse section through the cerebellum. *igl* = inner granular layer; *mo* = molecular layer; *egl* = outer granular layer; *Pl* = Purkinje layer; *v* = fourth ventricle. *Bar* = 100 µm. **h** Aspect of the outer granular layer of the cerebellum. Some granule cells are densely labeled. *Bar* = 10 µm. *Arrows* point to autoradiographically labeled cells.

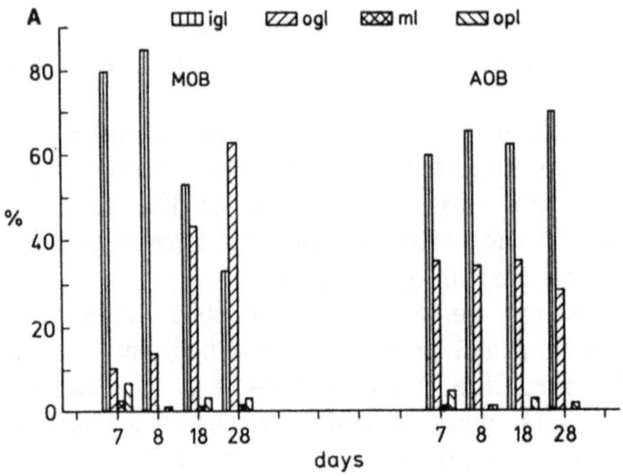

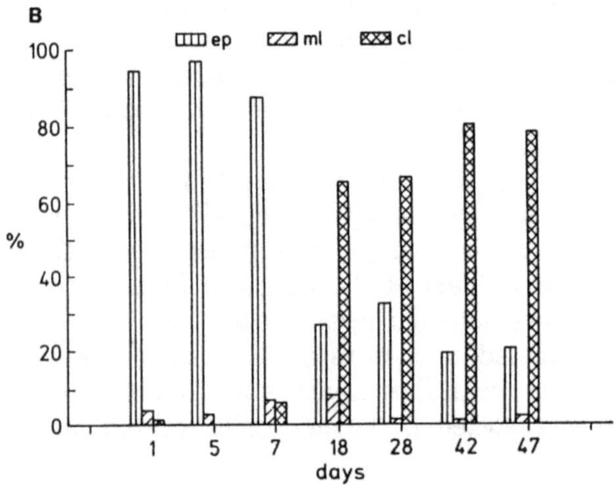

Fig. 2a-d. Distribution of autoradiographically labeled cells in the different layers (vertical axis = percentage values) after different survival times (horizontal axis = days after the last ^{3}H-thymidine injection). **a** Main (*MOB*) and accessory (*AOB*) olfactory bulbs. Observe the shift of labeled cells from the inner (*igl*) to the outer (*ogl*) parts of the granular layer in the MOB with increasing survival time. *ml* = mitral layer; *opl* = outer plexiform layer. **b** *Nucleus sphericus*. Note the shift from the ependyma (*ep*) to the mural cell layer (*cl*) with increasing survival time. *ml* = marginal layer. **c** Medial cortex. Note the shift from ependyma (*ep*) to granular cell layer (*gl*) with increasing survival time. *ipl* = inner plexiform layer. **d** Cerebellum. Note the shift from external granular layer (*egl*) to internal granular layer (*igl*) with increasing survival time. *mo* = molecular layer; *Pl* = Purkinje layer.

To summarize, as a general rule, cell proliferation occurred in ependymal germinal centers, the newly formed cells being recruited onto granular layers after different migration times.

C

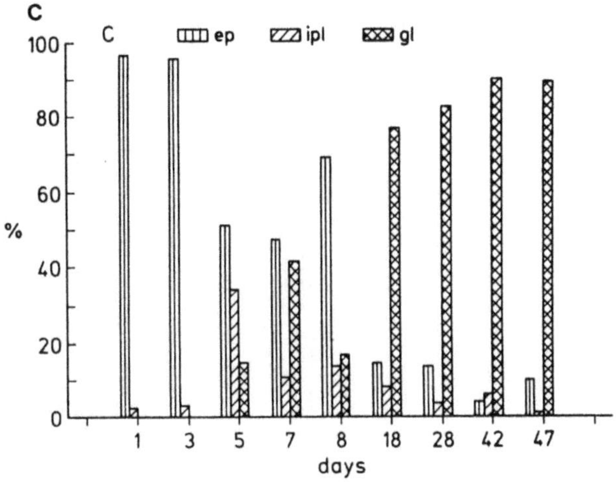

D

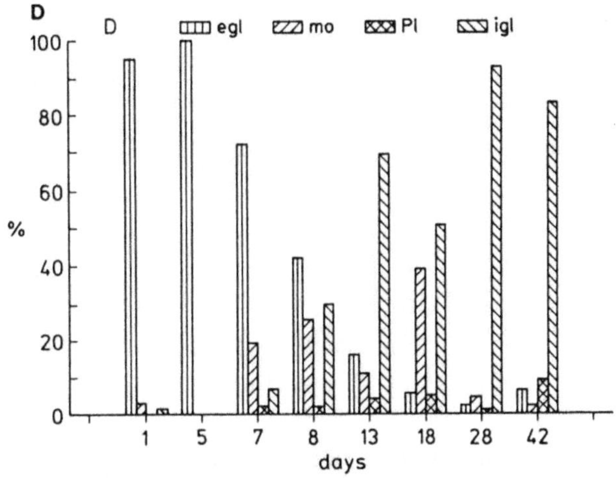

Autoradiography-HRP Double Labeling

In previous experiments, we showed that HRP injections in the outer plexiform layer of the dorsal cortex of *Podarcis hispanica* labeled cell bodies in the medial cortex (i.e. those located in the dorsalmost part of the principal cell layer; Lopez-Garcia and Martinez-Guijarro 1988). Nuclei of cells in the same location showed accumulation of radioactive label in the ^{3}H-thymidine experiments described above, suggesting that migrating cells in the medial cortex may participate in intracortical projections to the dorsal cortex.

In autoradiographic-HRP double labeling experiments, a few neurons in the medial cortex of animals that received HRP injections in the dorsal cortex were simultaneously labeled with

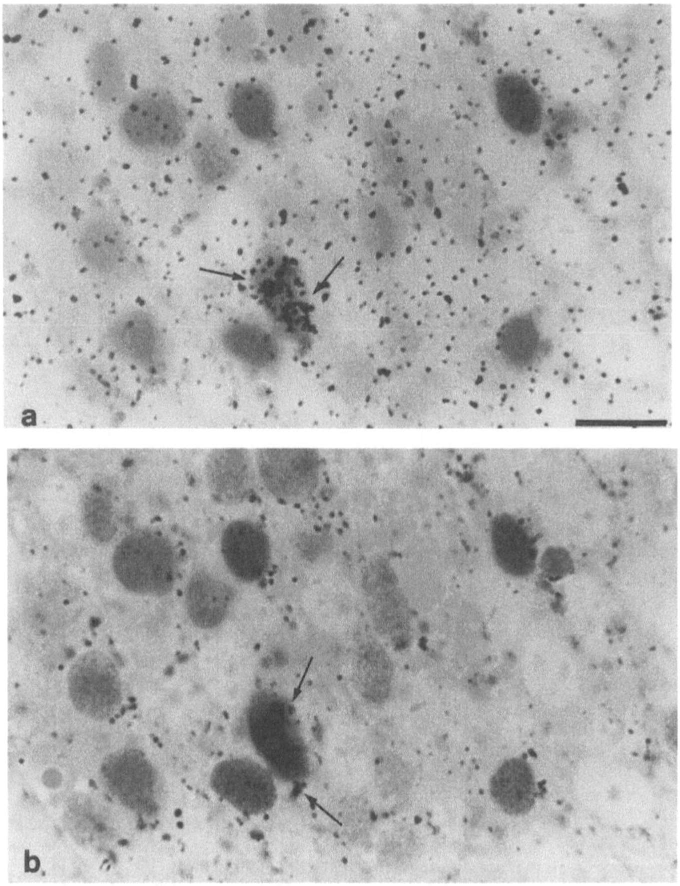

Fig. 3a,b. Double labeling. The same field photographed focusing on the autoradiographic emulsion (**a**) and the semithin section under it (**b**). The section shows part of the granule cell layer of a specimen that received an injection of HRP in the dorsal cortex two months after the last ^{3}H-thymidine injection. **a** *Arrows* point to a heavily labeled granule cell. **b** Some granule cells contain HRP in their cytoplasm. One of the cells (*arrows*) is shown to be autoradiographically labeled in **a**. *Bar* = 20 μm.

HRP and ^{3}H-thymidine (Fig. 3). This is evidence that newly formed neurons in the medial cortex send axons to the dorsal cortex and possibly make synaptic contacts there.

Discussion

1. Cells born in the brain of adult lizards are neurons
Electron microscopic studies have demonstrated that newly formed cells in the medial cortex and in the *nucleus sphericus* of lizards are neurons (Lopez-Garcia et al. 1988a). On this basis, we assume that newly formed cells in the olfactory bulbs and cerebellum are neurons as well.

2. Postnatally generated neurons send out axons
The ^{3}H-thymidine/HRP labeled somata observed in the medial cortex are evidence that at least some newly formed neurons send axons to the correct target places. Whether or not they can make synaptic contacts there remains an open question. A direct experimental confirmation of this matter seems difficult to obtain, yet indirect evidence strongly supports this possibility. Areas that receive projections from the medial cortex in lizards undergo intense synaptogenesis (our unpublished observations). Neurons at the medial cortex are recruited at the same rate that the number of synaptic contacts in its projection zones increases, probably due to an inflow of axons from newly formed neurons in those projection zones (Lopez-Garcia et al. 1988b).

3. Centers with postnatal neurogenesis are Timm reactive
The telencephalic structures with postnatal neurogenesis show a common histochemical feature: all of them display Timm-reactive zones (i.e. juxtasomatic laminae of the dorsal and dorsomedial cortices, marginal zone of the *nucleus sphericus* and inner plexiform layer of the olfactory bulbs; Molowny and Lopez-Garcia 1978; Molowny et al. 1987).

In the cerebral cortex of lizards, Timm-reactive zones are targets for axons coming from neurons located in the cell layer of the medial cortex (Lopez-Garcia and Martinez-Guijarro 1988). These axons give rise to Timm-positive presynaptic boutons. Some of the synaptic vesicles inside Timm positive-boutons contain zinc, presumably responsible for the pattern of Timm staining (Lopez-Garcia et al. 1983a,b; Martinez-Guijarro et al. 1987). Although it is possible that zinc functions in synaptic transmission, its role is still unclear.

4. Comparative aspects of postnatal neurogenesis
Delayed postnatal neurogenesis has been extensively studied in the retina and optic tectum of adult fish and amphibians (Straznicky

and Gaze 1971, 1972; Johns and Easter 1977; Stevenson and Yoon 1980; Johns 1982). Postnatal neurogenesis has also been detected in cultures of fish spinal cord (Leonard et al. 1978; Birse et al. 1980; Anderson and Waxman 1985).

In the mammalian brain, structures displaying postnatal neurogenesis are the olfactory bulbs, the hippocampal *fascia dentata* and the granule cell layer of the cerebellum. Their homology with corresponding structures in the lacertilian brain is well established. Thus, the olfactory bulbs and cerebella of mammals and reptiles are homologous structures whereas, the medial cortex of lizards is homologous to the mammalian *fascia dentata* (e.g. Lopez-Garcia et al. 1983b; Martinez-Guijarro et al. 1984; Desan 1988). In lizards, as in mammals, all of these structures exhibit postnatal neurogenesis. Only the *nucleus sphericus*, which also grows postnatally in lizards, lacks a mammalian counterpart. Recently, a few postnatally generated neurons have also been detected in the mammalian hypothalamus (Seress 1985).

Nucleus sphericus. The *nucleus sphericus*, particularly well developed in snakes and some lizards, receives a massive axonal projection from the ipsilateral accessory olfactory bulb (Heimer 1969; Halpern 1976) and, in turn, sends axons to the hypothalamus (Halpern and Silfen 1974). The *nucleus sphericus* is part of the archistriatum, traditionally considered as homologous to parts of the mammalian amygdala (Curwen 1939).

So far, the only other "striatal" center displaying postnatal neurogenesis is the "singing center" of canaries and other songbirds. There, postnatal neurogenesis is hormonally regulated (Nottebohm 1980; Goldman and Nottebohm 1983) and the newly formed neurons are incorporated into functional neural circuits as local interneurons (Burd and Nottebohm 1985; Paton et al. 1985).

Olfactory bulbs. Postnatal neurogenesis, especially during the 3 weeks following birth, has been previously described in the mammalian olfactory bulbs. As in lizards, most radioactively labeled cells in the olfactory bulbs of mammals were granular cells (Hinds 1968a,b; Altman 1969; Bayer 1983). In mammals, postnatal neurogenesis is more intense in the main than in the accessory olfactory bulbs (Altman 1969; Bayer 1983).

In the olfactory bulbs of lizards, ^{3}H-thymidine labeled nuclei move dorsally within the granular cell layer with increasing survival times. A similar transition, from inner to outer locations, has not been reported for the mammalian olfactory bulbs.

Medial Cortex. The medial cortex of lizards has been compared with the *fascia dentata* of the mammalian hippocampus on the basis of their similar cytoarchitecture, ontogenesis and histochemistry (Molowny and Lopez-Garcia 1978; Northcutt 1981; Desan 1988; Lopez-Garcia et al. 1988a,b). Both centers have most of their neuronal somata closely packed into a principal cell layer (granular layer), and their dendritic trees form an almost cell-free plexiform layer where the axonal inputs are highly stratified (Ramón y Cajal 1911; Lorente de Nó 1934; Blackstad and Kjaerheim 1961; Ebbesson and Voneida 1969; Regidor et al. 1974; Lacey 1978; Eckenhoff and Rakic 1984; Berbel et al. 1987).

Cells in the granular layer of reptiles and mammals send axons that synapse on the proximal dendrites of pyramidal (bipyramidal in reptiles) neurons (Ramón y Cajal 1911; Lorente de Nó 1934; Martinez-Guijarro 1984). The terminals of cells of the granular layer display an intense Timm reactivity, indicative of high zinc contents (Haug 1967; Geneser-Jensen et al. 1974; Molowny and Lopez-Garcia 1978; Lopez-Garcia et al. 1983b; Danscher 1984; Martinez-Guijarro et al. 1987; Molowny et al. 1987; Lopez-Garcia and Martinez-Guijarro 1988). Zinc seems to accumulate inside some synaptic vesicles of Timm-positive presynaptic boutons (Haug 1967; Lopez-Garcia et al. 1983b, 1984a; Perez-Clausell and Danscher 1985; Martinez-Guijarro et al. 1987) where it plays an unknown role in synaptic transmission (Euler 1962). The zinc-rich boutons of the dorsal cortex of lizards have been homologized to the mossy fiber boutons of the mammalian hippocampus (Molowny and Lopez-Garcia 1978; Lopez-Garcia et al. 1983b, 1984a; Martinez-Guijarro et al. 1984).

Strong support for the homology between the medial cortex of lizards and the mammalian *fascia dentata* comes from their late neurogenesis (Kirsche 1967; Hetzel 1974; Bayer et al. 1982; Kaplan and Bell 1983; Lopez-Garcia et al. 1984b, 1988a). In rats, the *fascia dentata* has extensive postnatal neurogenesis, acquiring up to 85% of its neurons during the 3 weeks following birth (Bayer 1980). Other mammals (e.g. primates) show an even more delayed postnatal neurogenesis of their *fascia dentata* (Rakic and Nowakowski 1981). In the case of lizards, postnatal neurogenesis in the medial cortex seems to extend into adulthood (Lopez-Garcia et al. 1984b, 1988a). The presence of apical shafts of ependymocytes that guide the migration of newly formed cells and the lack of any inside-outside gradient further support the idea of either a convergence or a common origin of the mechanisms regulating the neurogenetical processes in these cortical regions (Eckenhoff and Rakic 1984; Garcia-Verdugo et al. 1986).

5. Delayed neurogenetical potential in lizards

In lizards, a substantial pool of ^{3}H-thymidine labeled nuclei persisted in the generative zones of animals of all ages. This indicates the existence of a matrix or germinal zone in the ependymal lining of the medial cortex, the *nucleus sphericus* and the olfactory bulbs as well as in the outer granular layer of the cerebellar cortex.

The persistence of this generative pool in lizards has a counterpart in the *fascia dentata* of rats which increases its neuronal number during juvenile periods of life (Bayer et al. 1982). However, postnatal neurogenesis is more intense and lasts longer in lizards. The neurogenetical potential of the medial cortex of lizards may be related to the regenerative capacity of its cerebral cortex (Del Grande and Minelli 1980). Further work is required to understand the extent and biological significance of postnatal neurogenesis in the brain of lizards.

Acknowledgements. This study is specially dedicated to Regina L.M. for her help. We thank Dr. E. Font for his suggestions and comments on an earlier version of this manuscript. This study has been supported in part by a grant from the Spanish FISSS.

References

Altman J (1963) Autoradiographic investigations of cell proliferation in the brains of rats and cats. Anat Rec 143:573-529

Altman J (1969) Autoradiographic and histological studies of postnatal neurogenesis. IV. Cell proliferation and migration in the anterior forebrain with special reference to persisting neurogenesis in the olfactory bulb. J Comp Neurol 137:433-457

Altman J (1972a) Postnatal development of the cerebellar cortex in the rat. I. The external germinal layer and the transitional molecular layer. J Comp Neurol 145:353-398

Altman J (1972b) Postnatal development of the cerebellar cortex in the rat. II. Phases in the maturation of Purkinje cells and the molecular layer. J Comp Neurol 145:399-464

Altman J (1972c) Postnatal development of the cerebellar cortex in the rat. III. Maturation of the components of the molecular layer. J Comp Neurol 145:465-514

Altman J, Das GD (1965) Autoradiographic and histological evidence of postnatal hippocampal neurogenesis in rats. J Comp Neurol 124:319-335

Anderson MJ, Waxman SG (1985) Neurogenesis in tissue cultures of adult teleost spinal cord. Dev Brain Res 20:203-212

Angevine JB (1965) Time of neuron origin in the hippocampal region. An autoradiographic study in the mouse. Exp Neurol 13 (Suppl 2):1-70

Bayer SA (1980) Development of the hippocampal region in the rat. I. Neurogenesis examined with ^{3}H-thymidine autoradiography. J Comp Neurol 190:87-114

Bayer SA (1983) ^{3}H-thymidine-radiographic studies of neurogenesis in the rat olfactory bulb. Exp Brain Res 50:329-340

Bayer SA, Yackel JW, Puri PS (1982) Neurons in the dentate gyrus granular layer substantially increase during juvenile and adult life. Science 216:890-892

Berbel PJ, Martinez-Guijarro FJ, Lopez-Garcia C (1987) Intrinsic organization of the medial cerebral cortex of the lizard *Lacerta pityusensis*: a Golgi study. J Morphol 194:275-286

Birse SC, Leonard RB, Coggeshall RE (1980) Neuronal increase in various areas of the nervous system of the guppy, *Lebistes*. J Comp Neurol 194:291-301

Blackstad TW, Kjaerheim A (1961) Special axon-dendritic synapses in the hippocampal cortex: electron and light microscopic studies on the layer of mossy fibers. J Comp Neurol 117:133-159

Burd DB, Nottebohm F (1985) Ultrastructural characterization of synaptic terminals formed on newly generated neurons in a song control nucleus of the adult canary forebrain. J Comp Neurol 240:143-152

Curwen AO (1939) The telencephalon of *Tupinambis nigropunctatus*. III. Amygdala. J Comp Neurol 71:613-636

Danscher G (1984) Do the Timm sulphide silver method and the selenium method demonstrate zinc in the brain? In: Frederickson CJ, Howell GA, Kasarskis EJ (eds) The neurobiology of zinc. Part A: Physiochemistry, anatomy and techniques. Liss, New York, pp 273-287

Del Grande P, Minelli G (1980) Response of the regenerating telencephalon of *Lacerta viridis* to nerve growth factor. Z Mikrosk Anat Forsch 94:785-793

Desan P (1988) Organization of the cerebral cortex in turtle. In: Schwerdtfeger WK, Smeets WJAJ (eds) The forebrain of reptiles. Karger, Basel, pp 1-11

Ebbesson SOE, Voneida TJ (1969) The cytoarchitecture of the pallium of the Tegu lizard (*Tupinambis nigropunctatus*). Brain Behav Evol 2:431-466

Eckenhoff MF, Rakic P (1984) Radial organization of the hippocampal dentate gyrus: A Golgi, ultrastructural and immunocytochemical analysis in the developing rhesus monkey. J Comp Neurol 223:1-21

Euler C von (1962) On the significance of the high zinc content in the hippocampal formation. In: Passouant P (ed) Physiologie de l'hippocampe. Colloque Int CNRS Montpellier 1961, CNRS, Paris, pp 135-145

Garcia-Verdugo JM, Farina I, Molowny A, Lopez-Garcia C (1986) The ultrastructure of putative migrating cells in the cerebral cortex of *Lacerta galloti*. J Morphol 189:189-199

Geneser-Jensen FA, Haug FMS, Danscher G (1974) Distribution of heavy metals in the hippocampal region of the guinea pig. A light microscope study with Timm's sulphide silver method. Z Zellforsch 147:441-478

Goffinet AM, Daumerie CH, Langerwerf B, Pieau C (1985) Neurogenesis in reptilian cortical structures: ^{3}H-thymidine autoradiographic analysis. J Comp Neurol 243:106-116

Goldman SA, Nottebohm F (1983) Neuronal production, migration and differentiation in a vocal control nucleus of the adult female canary brain. Proc Natl Acad Sci USA 80:2390-2394

Halpern M (1976) The efferent connections of the olfactory bulb in the snakes *Thamnophis sirtalis* and *Thamnophis radix*. J Morphol 150:553-578

Halpern M, Silfen R (1974) The efferent connections of the *nucleus sphericus* in the snake *Thamnophis sirtalis*. Anat Rec 178:378

Haug FMS (1967) Electron microscopical localization of the zinc in hippocampal mossy fiber synapses by a modified silver procedure. Histochemie 8:355-368

Heimer L (1969) The secondary olfactory connections in mammals, reptiles and sharks. Ann N Y Acad Sci 167:129-146

Hetzel W (1974) Die Ontogenese des Telencephalons bei *Lacerta sicula* (Rafinesque) mit besonderer Berücksichtigung der pallialen Entwicklung. Zool Beitr 20:361-458

Hinds JW (1968a) Autoradiographic study of histogenesis in the mouse olfactory bulb. I. Time of origin of neurons and neuroglia. J Comp Neurol 134:287-304

Hinds JW (1968b) Autoradiographic study of histogenesis in the mouse olfactory bulb. II. Cell proliferation and migration. J Comp Neurol 134:305-322

Johns PR (1982) Formation of photoreceptors in larval and adult goldfish. J Neurosci 2:178-198

Johns PR, Easter SS (1977) Growth of the adult goldfish eye. II. Increase in retinal cell number. J Comp Neurol 176:331-342

Kaplan MS, Bell DH (1983) Neuronal proliferation in the 9-month-old rodent: autoradiography study of granule cells in the hippocampus. Exp Brain Res 52:1-5

Kaplan MS, Hinds JW (1977) Neurogenesis in the adult rat: electron microscopic analysis of light radiographs. Science 197:1092-1094

Kirsche W (1967) Über postembryonale Matrix-Zonen im Gehirn verschiedener Vertebraten und deren Beziehung zur Hirnbauplanlehre. Z Mikrosk Anat Forsch 76:313-406

Lacey DC (1978) The organization of the hippocampus of the fence lizard: a light microscopic study. J Comp Neurol 182:247-264

Leonard RB, Coggeshall RE, Willis WD (1978) A documentation of an age related increase of neuronal and axonal numbers in the stingray. J Comp Neurol 179:13-22

Lopez-Garcia C, Martinez-Guijarro FJ (1988) Neurons in the medial cortex give rise to Timm-positive boutons in the cerebral cortex of lizards. Brain Res 463:205-217

Lopez-Garcia C, Molowny A, Perez-Clausell J (1983a) Volumetric and densitometric study in the cerebral cortex and septum of a lizard (*Lacerta galloti*) using the Timm-method. Neurosci Lett 40:13-18

Lopez-Garcia C, Soriano E, Molowny A, Garcia-Verdugo JM, Berbel PJ, Regidor J (1983b) The Timm-positive system of axonic terminals of the cerebral cortex of *Lacerta*. In: Grisolia S, Guerri C, Samson F, Norton S, Reinoso-Suarez F (eds) Ramon y Cajal's contribution to the neurosciences. Elsevier, Amsterdam, pp 137-148

Lopez-Garcia C, Molowny A, Perez-Clausell J, Martinez-Guijarro FJ (1984a) A sulphide-osmium procedure for detection of metal-containing synaptic boutons in the lizard cerebral cortex. J Neurosci Methods 11:211-220

Lopez-Garcia C, Tineo P, Del Corral J (1984b) Increase of the neuron number in some cerebral cortical areas of a lizard, *Podarcis hispanica* (Steind., 1870), during postnatal periods of life. J Hirnforsch 25:255-259

Lopez-Garcia C, Molowny A, Garcia-Verdugo JM, Ferrer I (1988a) Delayed postnatal neurogenesis in the cerebral cortex of lizards. Developm Brain Res 43:167-174

Lopez-Garcia C, Molowny A, Rodriguez-Serna R, Garcia-Verdugo JM, Martinez-Guijarro FJ (1988b) Postnatal development of neurons in the telencephalic cortex of lizards. In: Schwerdtfeger WK, Smeets WJAJ (eds) The forebrain of reptiles. Karger, Basel, pp 122-130

Lorente de Nó R (1934) Studies on the structure of the cerebral cortex. II. Continuation of the study of the ammonic system. J Psychol Neurol 46:113-177

Martinez-Guijarro FJ, Berbel PJ, Molowny A, Lopez-Garcia C (1984) Apical dendritic spines and axonic terminals in the bipyramidal neurons of the dorsomedial cortex of lizards (*Lacerta*). Anat Embryol 170:321-326

Martinez-Guijarro FJ, Molowny A, Lopez-Garcia C (1987) Timm-staining is correlated with the density of Timm-positive presynaptic structures in the cerebral cortex of lizards. Histochemistry 86:315-319

Molowny A, Lopez-Garcia C (1978) Estudio citoarquitectonico de la corteza cerebral de reptiles. III: localizacion histoquimica de metales pesados y definicion de subregiones Timm positivas en la corteza de *Lacerta, Chalcides, Tarentola* y *Malpolon*. Trab Inst Cajal Invest Biol 70:55-74

Molowny A, Martinez-Calatayud J, Juan MJ, Martinez-Guijarro FJ, Lopez-Garcia C (1987) Zinc accumulation in the telencephalon of lizards. Histochemistry 86:311-314

Northcutt RG (1981) Evolution of the telencephalon in nonmammals. Ann Rev Neurosci 4:301-350

Nottebohm F (1980) Testosterone triggers growth of brain vocal control nuclei in adult female canaries. Brain Res 189:429-436

Paton JA, O'Loughlin BE, Nottebohm F (1985) Cells born in adult canary forebrain are local interneurons. J Neurosci 5:3088-3093

Pérez-Clausell J, Danscher G (1985) Intravesicular localization of zinc in rat telencephalic boutons. A histochemical study. Brain Res 337:91-98

Rakic P (1971) Neuron - glia relationship during granule cell migration in developing cerebellar cortex: a Golgi and electron microscopic study in *Macaca rhesus*. J Comp Neurol 141:283-312

Rakic P, Nowakowski RS (1981) The time of origin of neurons in the hippocampal region of the rhesus monkey. J Comp Neurol 196:99-128

Ramón y Cajal S (1911) Histologie du systeme nerveux de l'homme et des vertebres. A. Maloine, Paris

Raymond PA, Easter SS (1983) Postembryonic growth of the optic tectum in goldfish. I. Location of germinal cells and numbers of neurons produced. J Neurosci 3:1077-1091

Regidor J, Martinez-Trujillo JM, Lopez-Garcia C, Marin F (1974) Estudio citoarquitectonico de la corteza cerebral de reptiles. II. Tipologia dendritica y distribucion neuronal en areas corticales de *Lacerta galloti*. Trab Inst Cajal Invest Biol 63:1-33

Rossellini-Austin L, Altman J (1979) The postnatal development of the main olfactory bulb of the rat. J Developm Physiol 1:295-313

Schlessinger AR, Cowan WM, Gottlieb DI (1975) An autoradiographic study of the time of origin and pattern of granule cell migration in the dentate gyrus of the rat. J Comp Neurol 159:149-176

Seress L (1985) Postnatal neurogenesis in the rat hypothalamus. Dev Brain Res 22:156-160

Stevenson J, Yoon MG (1980) Kinetics of cell proliferation in the halved tectum of adult goldfish. Brain Res 184:11-22

Straznicki K, Gaze RM (1971) The growth of the retina in *Xenopus laevis*: an autoradiographic study. J Embryol Exp Morphol 26:67-79

Straznicki K, Gaze RM (1972) Development of the optic tectum in *Xenopus laevis*: an autoradiographic study. J Embryol Exp Morphol 28:87-115

Neuronal Organization of the Striatum in the Alligator, *Alligator mississippiensis*

Philip S. Ulinski
Department of Organismal Biology and Anatomy,
The University of Chicago, Chicago, Illinois 60637, USA

The cerebral hemispheres of tetrapod vertebrates consist of three components. Each hemisphere has a pallium extending above its lateral ventricle. The pallium of mammals is formed entirely by a laminated cortex. However, the pallium of reptiles and birds includes a structure -- the dorsal ventricular ridge (DVR) -- that protrudes into the lateral ventricle and lacks distinct lamination (Ulinski 1983), as well as a laminated cortex (Ulinski 1990). The second component, the septum, forms the ventromedial wall of the hemisphere and is part of the limbic system. Third is the striatum, which forms the ventrolateral wall of the hemisphere and is named so because fibers running through the basal forebrain give it a striated appearance in sections stained for myelin. The striatum is not large in amphibians, reptiles and birds, but protrudes dorsally into the lateral ventricles of mammals forming the "ganglionic eminences" or basal ganglia.

A role for the striatum in organizing complex motor behaviors was suggested by clinical observations and has been amply confirmed by behavioral and neurophysiological work in mammals and other tetrapod vertebrates (e.g. Bugbee and Hodos 1979; Brauth 1987; Greenberg et al. 1988). Application of axon tracing techniques demonstrates that the striatum of mammals has distinct relationships with other regions of the brain and establishes that the pattern of striatal connections in reptiles and birds bears a general resemblance to that of mammals (e.g. Karten and Dubbeldam 1973; Reiner et al. 1984; Parent 1986). The pallium (cortex/DVR or both) projects to the striatum; the striatum is the source of subsequent projections to brainstem structures that ultimately modulate the activity of the optic tectum (or superior colliculus). A particular feature of the forebrain of mammals is the presence of extensive projections from the striatum to the thalamic nuclei projecting to motor cortex. I have suggested (Ulinski 1986)

Experimental Brain Research Series 19

that these projections evolved in concert with reorganizations of the limbs and jaws of early mammals that demanded the evolution of sophisticated motor control systems. Reciprocal connections between the striatum and dopaminergic cell groups (the substantia nigra and ventral tegmental area) in the midbrain tegmentum are also common to reptiles, birds and mammals (e.g. Brauth 1988a; Smeets 1988).

Although there has been progress in defining a common pattern of striatal connections, understanding the internal organization of the striatum has proven difficult. Early anatomists recognized several cytoarchitectonic subdivisions of the striatum in mammals and Ariens Kappers (e.g. Ariens Kappers et al. 1936) introduced a nomenclature for the components of the striatum (archistriatum, paleostriatum and neostriatum) that placed these into an evolutionary hierarchy. In essence, he proposed a linear model of striatal evolution in which the paleostriatum and neostriatum were added sequentially to the archistriatum throughout the progress of vertebrate evolution. From a functional viewpoint, this concept led to a serial model of information processing in which information from the pallium is relayed first to the neostriatum, then to the paleostriatum and ultimately to the brainstem. Motor behavior in mammals was seen as controlled by motor cortex, reptiles were seen as having only a rudimentary modulation of motor activity by the forebrain and birds were thought to show a hypertrophy of striatal structures (e.g. the hyperstriatum of birds) reflected in their largely instinctive behavioral repertoire.

A number of problems occur when one tries to identify the neostriatum and paleostriatum in a variety of vertebrates. One is that there was nothing linear about the evolution of the vertebrates, so a linear model of telencephalic evolution is inconsistent with the history of the vertebrates (Northcutt 1981). The behavioral characterizations of reptiles and birds, implicit in the earlier discussions, is now also recognized as naive. Furthermore, a careful reading of the literature indicates that terms such as "paleostriatum", "nucleus accumbens", "ventral pallidum" etc., are often used in different ways in various studies and sometimes within the same paper. Finally, recent studies addressing the neuronal organization of the striatum in mammals are revealing a very complicated internal structure that varies considerably between species of mammals (e.g. Parent 1986), raising questions about the generality of some classical concepts of striatal organization. It is thus not yet clear if the linear model of striatal organization can be validly extended beyond the mammals to other groups of vertebrates.

One limitation to progress is the absence of information on the morphology of striatal neurons for vertebrates other than mammals. The morphology of neurons in the striatum of mammals has been established using both classical and modern methods (e.g. Heimer et al. 1985), and details of the projections and neurotransmitter profiles of specific populations of neurons have become increasingly important in understanding the striatum of mammals (e.g. Graybiel and Ragsdale 1983). Since there are only a handful of observations on striatal neurons in reptiles (Ramon 1896; Crosby 1917; Northcutt 1967, 1970), Eva Vyas and I (Vyas and Ulinski 1988) chose to study neurons in the striatum of the American alligator, *Alligator mississippiensis*, because such information would compliment the considerable body of literature currently available on striatal connections and histochemistry in alligators. Our studies lead to a series of focused questions concerning striatal organization in reptiles, which are considered in the last section of this chapter.

Morphology of Striatal Neurons in Alligators[1]

The striatum in *Alligator* (Fig. 1) corresponds approximately to the ventrolateral area (defined by Crosby 1917). It is separated by a cell poor zone from the intermediolateral and dorsolateral areas, which lie within the anterior dorsal ventricular ridge (Clark and Ulinski 1984). Neurons in the striatum fall into two distinct groups in Nissl preparations. The most plentiful are medium sized neurons with somata measuring about 10 µm x 15 µm. They have little cytoplasm and lack distinct Nissl bodies, but do show a tendency to form clusters of neurons with apposed somata. Medium sized neurons are distributed throughout the striatum, but are relatively infrequent within the caudal, ventral part of the striatum which contains the dorsal peduncle of the lateral forebrain bundle. This bundle contains

---

[1] Neuroanatomical work on crocodilian reptiles has been done almost exclusively on two genera, the caimans of the genus ***Caiman*** and the American alligator of the genus ***Alligator***. These animals are closely related and placed by systematists in the family Alligatoridae. It is probably reasonable to combine data from the two genera in the context of the present, general discussion on striatal organization, and I will refer to the caimans plus ***Alligator*** as "alligators".

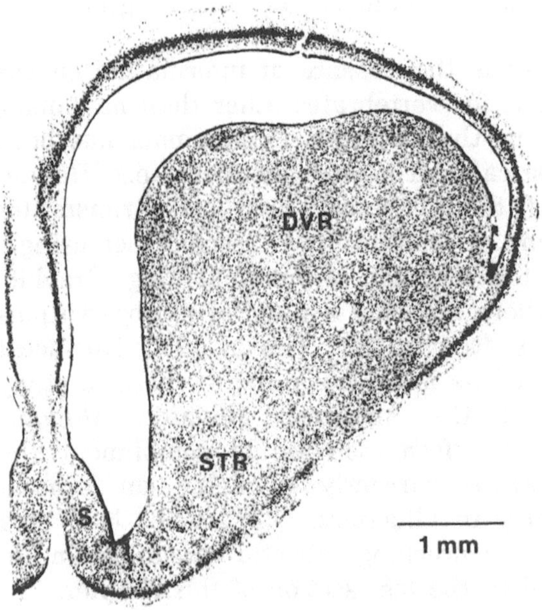

Fig. 1. Striatum in alligator. This photomicrograph illustrates the cytoarchitecture of the striatum and related structures in a transverse section through the rostral telencephalon of the alligator. The striatum (*STR*) is separated from the overlaying dorsal ventricular ridge (*DVR*) by a cell poor area. It blends medially with the septum (*S*) and is surrounded externally by olfactorecipient structures which are distinguished by a layer of densely packed, basophilic cells.

axons running from sensory thalamic nuclei, through the striatum and into the dorsal ventricular ridge to terminate in modality-specific sensory areas (Pritz 1974, 1975; Pritz and Northcutt 1980). The second group of neurons have large, polygonal somata that measure about 15 µm x 25 µm and contain extensive cytoplasm and distinct Nissl bodies. They are most frequent within the fibers of the dorsal peduncle of the lateral forebrain bundle, but occasionally occur in the more rostral regions of the striatum. The striatum thus consists of two regions with indistinct boundaries. A caudal region is characterized by large cells associated with the lateral forebrain bundle and corresponds to the ventro caudal region of the ventrolateral area of Brauth and Kitt (1980). It is surrounded medially, dorsally, rostrally and laterally by a rostral region characterized by medium sized neurons.

Striatal neurons can also be divided into two major populations in Golgi preparations. The most striking have large somata and

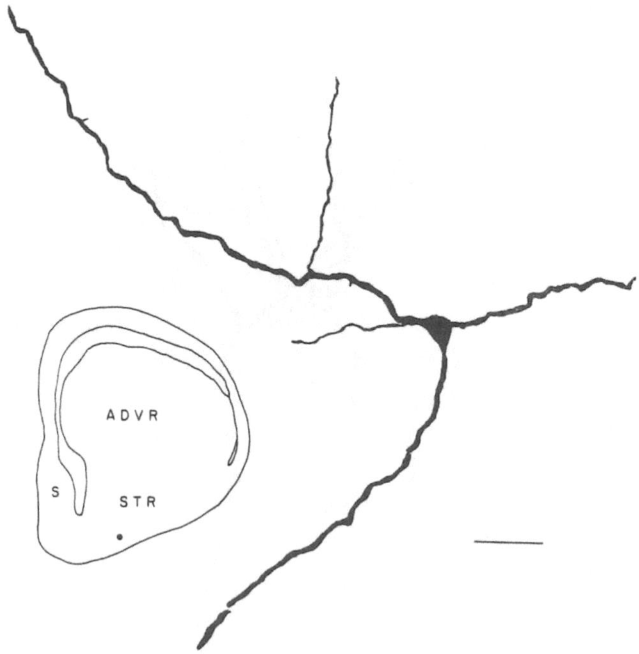

Fig. 2. Large neuron. This is a drawing of a large neuron with smooth dendrites. The inset shows a transverse section of the telencephalon with the location of the soma of this particular neuron indicated by the *dot*. It was situated in the ventral striatum just internal to the olfactorecipient structures. *Scale bar*, 50 μm. *ADVR*, anterior dorsal ventricular ridge; *STR*, striatum; *S*, septum.

extensive dendritic trees. They clearly correspond to the large neurons of the caudal region (Fig. 2). Each of these cells has two or three thick and smooth dendrites that extend, with relatively little branching, in roughly the transverse plane. Comparisons of the dimensions of the dendritic trees of these cells in brains cut in different planes of section indicate that each neuron can be contained within a rectangular parallelepiped measuring about 1000 μm along its medial-lateral and dorsal-ventral axes and 150 μm along its rostral-caudal axis. These cells stain positively with antisera prepared against the peptide lysine-asparagine-neurotensin (LANT6), which Reiner and Carraway (1985) hypothesize may colocalize with GABA. The large cells can be contrasted to three subpopulations of neurons with medium sized somata. The most frequent have roughly spherical dendritic trees with diameters of about 300 μm (Fig. 3) and are composed of three to five very spiny dendrites. Medium sized neurons with sparse,

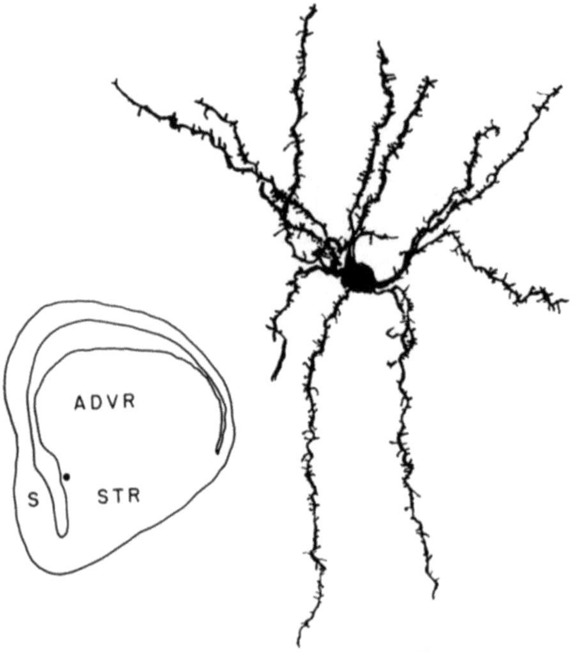

Fig. 3. Medium sized spiny neuron. This is a drawing of a medium sized neuron with very spiny dendrites from a Golgi preparation. The inset shows a transverse section of the telencephalon with the location of the soma of this particular neuron indicated by the *dot*. It was situated in the medial striatum (*STR*), just below its border with the anterior dorsal ventricular ridge (*ADVR*). *S*, septum. *Scale bar*, 50 µm.

spiny dendrites or those with smooth, beaded dendrites are also present, but relatively rare. Immunocytochemical data indicate medium sized striatal neurons stain with antisera prepared against choline acetyltransferase (Brauth et al. 1985), met- and leu-enkephalin (Brauth 1984), substance P (Brauth et al. 1983) and calcium binding protein (Brauth et al. 1988). However, these studies do not provide enough morphological detail to associate each of the morphological groups of cells with distinct neurotransmitter profiles.

Both immunocytochemical and Golgi data show that the various neuronal populations are spread throughout the striatum in a series of spatial gradients, rather than being segregated sharply into cytoarchitectonic units. Somata staining positively with antisera against substance P are frequent medially and dorsally in the rostral region and circumscribe the large cells in the caudal striatum (Brauth et al. 1983). Somata stained with antisera against enkephalin show the same general pattern, but with a greater tendency for enkephalin-positive cells to occur laterally in the rostral region (Brauth 1984). By contrast, somata stained with antisera against choline acetyltransferase tend to be embedded in the lateral forebrain bundle (Brauth et al. 1985). They are smaller (10 - 12 μm) in the rostral region, but larger (12 - 15 μm) in the caudal striatum. Somata stained with antiserum against calcium binding protein are frequent in the dorsal and lateral parts of the rostral region (Brauth 1988b; Brauth et al. 1988). We (Vyas and Ulinski 1988) plotted the spatial distribution of 963 neurons in three brains prepared by Golgi methods and found that all four morphological types of neurons occur throughout the striatum. As in the Nissl material, large neurons occur more frequently in the caudal region.

Striatal Connections in Alligators

The striatum of alligators receives two major sets of afferent projections. The first originate from neurons in the anterior dorsal ventricular ridge (ADVR; Schapiro 1964; Brauth and Kitt 1980). The organization of these projections has not been studied in detail in alligators, but Nautiyal and Ulinski (1986) used the anterograde transport of HRP to study the organization of the DVR-striatal system in turtles. They found tha the projection is topographically organized, with neurons in the medial part of ADVR descending through the medial part of the striatum and those in the lateral part of ADVR descending through the lateral part of the striatum. Each axon consists of a smooth parent branch that bears many varicose collaterals extending from its shaft at oblique angles. The spatial relationships are such that each axon can potentially synapse upon both large and medium sized populations of neurons through much of the striatum. Since auditory, somatosensory and visual regions are present in the medial, central and lateral parts of the ADVR of alligators, respectively (Pritz 1974, 1975; Pritz and Northcutt 1980), these data predict that many striatal neurons receive convergent auditory, somatosensory and visual information.

The second major afferent system arises from two cell groups in the midbrain tegmentum and is continuous laterally with the substantia nigra (which was called the nucleus penduncolopontinus tegmenti in the earlier literature; see Brauth 1988a). These two groups contain many neurons that stain positively with antisera against the synthetic enzyme tyrosine hydroxylase, suggesting they are dopaminergic, and a smaller number of neurons that stain with antisera against serotonin (Brauth et al. 1983; Brauth 1988a). They project topographically to the striatum, with the ventral tegmental area projecting to the most medial parts of the striatum, including the wing of the striatum that extends ventral to the recess of the lateral ventricle, and the substantia nigra projecting to the more lateral parts of the striatum (Brauth and Kitt 1980; Brauth 1988b). The entire striatum stains positively with antisera against tyrosine hydroxylase (Brauth 1988b), but the density of immunoreactive fibers is lower laterally than medially. Similarly, the regions of the striatum surrounding the lateral forebrain bundle stain positively with antisera against serotonin (Brauth 1988b). At first glance, this suggests a differential distribution of dopaminergic and serotonergic inputs onto the various populations of neurons within the striatum, but the extensive dendritic fields of the large cells make it possible for them to receive dopaminergic and serotonergic inputs on those parts of their dendritic fields extending beyond the lateral forebrain bundle. All four morphological groups of neurons may then receive inputs from both the ADVR and the midbrain dopaminergic and serotonergic groups.

By contrast, several lines of evidence suggest that the various morphological groups of striatal neurons participate in distinct efferent channels. Autoradiographic tracing studies (Reiner et al. 1980; Brauth 1988b) demonstrate that the striatum as a whole projects to a number of brainstem structures, including the habenula, locus coeruleus and neighboring cell groups and ventral thalamic groups situated in the ventral peduncle of the lateral forebrain bundle. It is convenient, however, to emphasize three major projections. The first consists of fibers that run caudally down the brainstem reaching into the lateral parts of the medullary reticular formation. Second are fibers that project to the ventral tegmental area and substantia nigra. Third are fibers that project to a pretectal structure known as the dorsal nucleus of the posterior commissure. A common feature of these populations is that they place the striatum in contact with premotor neurons in the brainstem reticular formation. The striatal-reticular projections do so directly, while those to the midbrain tegmentum and pretectum do so indirectly because both the substantia nigra and the nucleus of the posterior commissure project to the optic tectum (Reiner et al. 1980), a source of extensive tectoreticular projections.

One hint of an organization in striatal efferent projections comes from studies in which injections of tritiated amino acids are made in restricted areas of the striatum. Brauth (1988b), for example, placed injections in the caudal region of the striatum and showed projections to the pontine and medullary reticular formation. Injections of HRP in the pontine reticular formation retrogradely labeled neurons in the caudal region of the striatum. Labeled cells lie within the region containing large neurons showing LANT6-like immunoreactivity (Reiner and Carraway 1985) and medium sized neurons showing choline acetyltransferase-like immunoreactivity (Brauth et al. 1985). A reasonable hypothesis is that the large striatal neurons are LANT6 positive (medium cholinergic striatal) neurons that project to the brainstem reticular formation. Brauth and Kitt (1980) placed injections in the more rostral regions of the striatum and showed projections to the other major striatal targets, including the ventral tegmental area, substantia nigra and nucleus of the posterior commissure. HRP injections in the substantia nigra retrogradely labeled neurons scattered throughout both the rostral and caudal regions of the striatum. Axon systems showing immunoreactivity to antisera against both substance P (Brauth et al. 1983) and the enkephalins (Brauth 1984) can be traced from the striatum to the ventral tegmental area and substantia nigra. Knife cuts caudal to the striatum lead to a depletion of substance P immunoreactivity in the midbrain tegmentum (Brauth et al. 1983). These observations suggest that substance P-ergic and enkephalinergic medium sized neurons, located throughout the striatum, project to the mesencephalic targets of the striatum, while medium sized neurons, with some still unknown transmitter(s), project to the nucleus of the posterior commissure.

Issues in Striatal Organization

Our data on the morphology of striatal neurons in *Alligator* can be combined with data on the connections and histochemistry of the striatum in alligators, obtained by Brauth, Reiner and their coworkers, to identify a series of specific issues, each of which can be resolved using modern neurobiological techniques. The issues are evident when comparing Figs. 4a and 4b. Figure 4a summarizes the organization of the striatum in alligators, based on Figs. 2 and 3 in Reiner et al. (1984). Figure 4b is a summary of striatal organization of the striatum in alligators, based on the previous two sections of this paper.

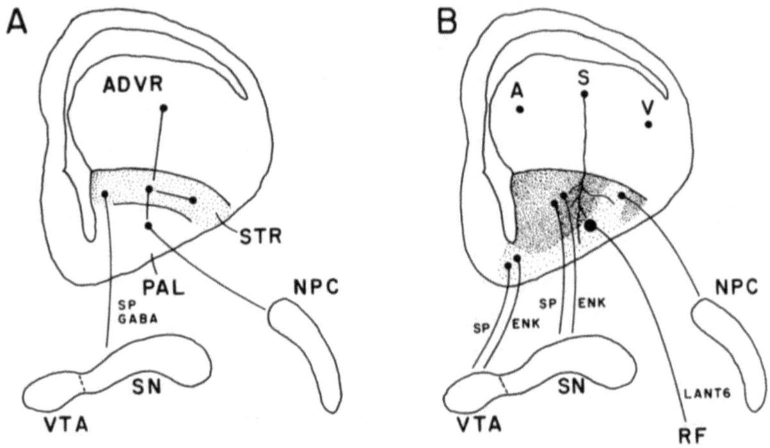

Fig. 4a,b. Models of striatal organization. This figure shows two different models of striatal organization. Figure **4a** is a synthesis of Figs. 2 and 3 from the review by Reiner et al. (1984). **A** Neurons in the pallium (*ADVR*) project to the "striatum" (*STR*). The "striatum" contains interneurons that form connections entirely within the "striatum". Other neurons project from the "striatum" to the "pallidum" or to the dopaminergic cell groups of the midbrain (substantia nigra, *SN*; ventral tegmental area, *VTA*). One set of neurons projects to the ventral tegmental area and substantia nigra while another projects to the "pallidum" (*PAL*). Projections to the midbrain use substance P (*SP*) or GABA as neurotransmitters. The "pallidum" projects to the nucleus of the posterior commissure (*NPC*). The "striatum" receives a dopaminergic projection from the midbrain dopaminergic groups which is represented by stipple. Figure **4b** is based on the data reviewed in this article. The anterior dorsal ventricular ridge contains auditory (*A*), somatosensory (*S*) and visual (*V*) areas. Each area has neurons whose axons descend into the striatum and bear collaterals. Medium sized and large neurons are spread throughout the striatum. Thus, both populations of neurons receive projections from neurons in ADVR. Medium sized neurons project to the midbrain dopaminergic groups and to the nucleus of the posterior commissure. Projections to the ventral tegmental area and substantia nigra use substance P and the enkephalins (*ENK*) as neurotransmitters. Large striatal neurons project to the brainstem reticular formation (*RF*) and contain LANT6. The origin of projections to the nucleus of the posterior commissure is not certain, but there is no indication it originates from large neurons. It is thus shown originating from medium sized neurons.

Striatal compartments The model in Fig. 4a recognizes two compartments in the striatum. One is a "striatum" that corresponds to the neostriatum (or caudate and putamen) of mammals, and the second is a "pallidum" that corresponds to the paleostriatum (or globus pallidus) of mammals. The "striatum" is seen as containing small neurons, the "pallidum" is seen as containing large neurons, and the "striatum" and "pallidum" each have distinct afferent and efferent connections. The model in Fig. 4b indicates that medium and large sized neurons are spread throughout the striatum in Nissl material and do not respect the boundaries of identifiable compartments. In particular, the dendrites of the large cells extend into the field usually identified as the "striatum". Afferents terminating in either the "striatum" or "pallidum" can synapse upon both medium and large sized neurons. Consequently, the striatum in Fig. 4b is not partitioned into distinct "striatal" and "pallidal" compartments.

Pallial-pallidal-striatal projections The model in Fig. 4a indicates that the DVR projects to the striatum and the striatum then projects to the pallidum. If our data from turtles (Nautiyal and Ulinski 1986) can be generalized to alligators, single neurons in the DVR have axons that bear varicose collaterals throughout the striatum (Fig. 4b). Thus, both medium and large sized neurons receive direct inputs from the auditory, somatosensory and visual areas of the ADVR. It is not yet clear if any medium sized neurons in the striatum project exclusively to the pallidum. The alternative is that medium sized neurons have axons that project from the striatum to the brainstem as well as collaterals that synapse upon other neurons within the striatum. The concept that striatum projects to pallidum is based on experiments in which autoradiographic tracers are injected in the striatum. However, none of these studies have been conducted in sufficient detail to determine if striatal neurons actually form synapses on pallidal neurons. The finding (Russchen et al. 1976; Russchen and Jonker 1988) that axons, labeled by injections of the anterograde tracer PHA-L into the striatum, bear varicosities as they course through the pallidum, but are smooth en route to the brainstem indicates that at least some striatal neurons synapse in the pallidum. However, it does not unequivocally distinguish between the two alternatives because it does not determine if individual axons both collateralize within the striatum and project to the brainstem.

Relation of the striatum to midbrain dopaminergic groups A series of midbrain structures, including the substantia nigra and ventral tegmental area, contain dopaminergic neurons that project to the striatum. The ventral tegmental area projects to the medial striatum, while the substantia nigra projects to the lateral striatum. The model in Fig. 4a suggests that the midbrain dopaminergic groups project to the striatum. However, immunocytochemical studies in alligators indicate that fibers showing tyrosine hydroxylase-like immunoreactivity are spread through the striatum (Fig. 4b). They are less dense laterally than medially, perhaps reflecting the compact arrangement of fibers of the dorsal peduncle of the lateral forebrain bundle in the lateral striatum. The dendritic field sizes of large neurons makes it likely that both medium and large sized neurons receive direct inputs from the dopaminergic cell groups of the midbrain. There is a surprising amount of interspecific variation between reptiles in the distribution of axons showing dopamine-like immunoreactivity in the striatum (e.g. Smeets 1988). The significance of this variation is not clear, but it does not fit the conventional distinction between striatum and pallidum in any obvious way.

The model in Fig. 4a indicates that the striatum is the source of a reciprocal projection to the dopaminergic cell groups of the midbrain. These projections include axons using substance P or GABA as neurotransmitters. The model in Fig. 4b indicates that the entire striatum gives rise to a topographically organized projection to the midbrain dopaminergic groups, with the medial striatum projecting to the ventral tegmental area and the lateral striatum projecting to the substantia nigra. Medium sized neurons containing both substance P and the enkephalins participate in the projection. It is not known if different subgroups of medium sized neurons use different neurotransmitters, but it is known that somata containing both transmitters occur throughout the striatum. There is no direct information on the distribution of GABA-containing neurons in the striatum or their projection to the brainstem.

Striatal projections to the nucleus of the posterior commissure The model in Fig. 4a indicates that the pallidum projects to the nucleus of the posterior commissure in reptiles and birds. This is seen as quantitatively different from the situation in mammals where the globus pallidus gives rise to a major projection to the thalamus. The model in Fig. 4b indicates that the large neurons of the striatum project to the brainstem reticular formation. These neurons have a distinctive morphology that can be identified in immunocytochemical studies which indicate they show LANT6-like or choline acetyltransferase-like immunoreactivity. The origin of the projection to the nucleus of the posterior commissure, or which neurotransmitter it uses, is not clear.

Topographic organization of striatal efferents The model in Fig. 4a indicates that the striatum and pallidum have distinct efferent projections. The model in Fig. 4b indicates that the efferent projections from the striatum to the midbrain dopaminergic cell groups is topographically organized, as mentioned above. It does not comment on the organization of striatal efferents to other structures. There is, however, an indication that these projections are organized in some fashion. Brauth (1988b) reports that the caudal part of the striatum (the ventral pallidum) projects to a series of brainstem structures, but not to the substantia nigra. Similarly, Russchen and Jonker (1988) report projections from different regions of the striatum to different structures in the brainstem in the lizard *Gekko gecko*.

Conclusion

Resolution of the issue discussed above will do much to clarify our concept of the organization of the striatum in reptiles. However, it will also provide the information necessary for detailed comparisons between striatal organization in reptiles and mammals. It will be particularly important to determine if the distinction between matrix and patch areas within the neostriatum of mammals (e.g. Heimer et al. 1985) extends to reptiles. There are now reports of inhomogeneities in the distribution of various compounds in the striatum of reptiles (e.g. Russchen et al. 1987a), but it is not at all clear if these inhomogeneities can be equated to those associated with the matrix and patch components of the neostriatum of mammals.

References

Ariens Kappers CU, Huber GC, Crosby EC (1936) The comparative anatomy of the nervous system of vertebrates, including man. Republished in 1967 by Hafner, New York.

Brauth SE (1984) Enkephalin-like immunoreactivity within the telencephalon of the reptile *Caiman crocodilus*. Neuroscience 11:345-358

Brauth SE (1987) Motor functions of the *Caiman* basal ganglia. Neurosci Abstr 13:978

Brauth SE (1988a) Catecholamine neurons in the brainstem of the reptile *Caiman crocodilus*. J Comp Neurol 270:313-326

Brauth SE (1988b) The organization and projections of the paleostriatal complex of *Caiman crocodilus*. In: Schwerdtfeger WK, Smeets WJAJ (eds) The forebrain of reptiles, Karger, Basel, pp 60-76

Brauth SE, Kitt CA (1980) The paleostriatal system of *Caiman crocodilus*. J Comp Neurol 189:437-465

Brauth SE, Kitt CA, Gerfen CR (1988) Calcium binding protein in the basal ganglia system in a non-mammalian vertebrate: an immunohistochemical study in the reptile *Caiman crocodilus*. Brain Res 452:367-372

Brauth SE, Kitt CA, Price DL, Wainer BH (1985) Cholinergic neurons in the telencephalon of the reptile *Caiman crocodilus*. Neurosci Lett 58:235-240

Brauth SE, Reiner A, Kitt CA, Karten HJ (1983) The substance P containing striatotegmental path in reptiles: an immunohistochemical study. J Comp Neurol 219:305-327

Bugbee NM, Hodos W (1979) The basal ganglia-tectal pathway: its role in visually guided behavior in the pigeon. Neurosci Abstr 5:68

Clark JM, Ulinski PS (1984) A Golgi study of anterior dorsal ventricular ridge in the alligator, *Alligator missisippiensis*. J Morphol 179:153-174

Crosby EC (1917) The forebrain of *Alligator mississippiensis*. J Comp Neurol 27:325-403

Graybiel AM, Ragsdale CW (1983) Biochemical anatomy of the striatum. In: Emson PC (ed) Chemical neuroanatomy. Raven, New York, pp 427-504

Greenberg N, Font E, Switzer III RC (1988) The reptilian striatum revisited: studies on *Anolis* lizards. In: Schwerdtfeger WK, Smeets WJAJ (eds) The forebrain of reptiles, Karger, Basel, pp 162-177

Heimer L, Alheid GF, Zaborsky L (1985) Basal ganglia. In: Paxinos G (ed) The rat nervous system, vol I. Academic, Sydney, pp 37-86

Karten HJ, Dubbeldam JL (1973) The organization and projections of the paleostriatal complex in the pigeon (*Columbia livia*). J Comp Neurol 148:61-89

Nautiyal J, Ulinski PS (1986) Morphology of efferents from dorsal ventricular ridge to striatum in turtles. Neurosci Abstr 12:1541

Northcutt RG (1967) Architectonic studies of the telencephalon of *Iguana iguana*. J Comp Neurol 130:107-148

Northcutt RG (1970) The telencephalon of the western painted turtle, *Chrysemis picta belli*. Univ Illinois Press, Urbana (Illinois Biological Monographs 43)

Northcutt RG (1981) Evolution of the telencephalon in nonmammals. Annu Rev Neurosci 4:301-350

Parent A (1986) Comparative neurobiology of the basal ganglia. Wiley, New York

Pritz MB (1974) Ascending connections of a midbrain auditory area in a crocodile *Caiman crocodilus*.J Comp Neurol 153:179-198

Pritz MB (1975) Anatomical identification of a telencephalic visual area in crocodiles: Ascending connections of nucleus rotundus in *Caiman crocodilus*. J Comp Neurol 164:323-338

Pritz MB, Northcutt RG (1980) Anatomical evidence for an ascending somatosensory pathway to the telencephalon in crocodiles, *Caiman crocodilus*. Exp Brain Res 40:342-345

Ramon P (1896) Estructura del encefalo del camaleon. Rev Trim Micrografica 1:146-182

Reiner A, Brauth SE, Karten HJ (1984) Evolution of the amniote basal ganglia. Trends Neurosci 7:320-325

Reiner A, Brauth SE, Kitt CA, Karten HJ (1980) Basal ganglionic pathways to the tectum: studies in reptiles. J Comp Neurol 193:565-589

Reiner A, Carraway RE (1985) Phylogenetic conservatism in the presence of a neurotensin-related hexapeptide in neurons of globus pallidus. Brain Res 340:357-362

Russchen FT, Jonker AJ (1988) Efferent connections of the striatum and nucleus accumbens in the lizard *Gekko gecko*. J Comp Neurol 276:61-80

Russchen FT, Smeets WJAJ, Hoogland PU (1987a) Histochemical identification of pallidal and striatal structures in the lizard *Gekko gecko*. Evidence for compartmentalization. J Comp Neurol 256:329-341

Russchen FT, Smeets WJAJ, Lohman AHM (1987b) On the basal ganglia of a reptile, the lizard *Gekko gecko*. In: Carpenter MB, Jayaraman A (eds) Advances in behavioral Biology, vol 32. The basal ganglia. Plenum, New York, pp 261-282

Schapiro H (1964) Motor functions and their anatomical basis in the forebrain and tectum of the alligator. Thesis, University of Florida

Smeets WJAJ (1988) The distribution of dopamine immunoreactivity in the forebrain and midbrain of the snake *Python regius*: a study with antibodies against dopamine. J Comp Neurol 271:115-129

Ulinski PS (1983) Dorsal ventricular ridge: A treatise on forebrain organization in reptiles and birds. Wiley, New York

Ulinski PS (1986) Neurobiology of the therapsid-mammal transition. In: Hotton N, Maclean PD, Roth JJ, Roth EC (eds) The ecology and biology of mammal-like reptiles. Smithsonian Institution Press, Washington, pp 149-172

Ulinski PS (1990) Cerebral cortex in reptiles. In Jones EG, Peters A (eds) Cerebral cortex, vol 8, Plenum, New York (in press)

Vyas E, Ulinski PS (1988) Neuronal organization of striatum in the *Alligator*: a Golgi study. J Neurosci Suppl 8:20

Cortical Architectonic Development: a Comparative Study in Reptiles

André M. Goffinet
Department of Physiology, FNDP Medical College 61,
Rue de Bruxelles, B-5000 Namur, Belgium

The cerebral cortex is the product of a long evolutionary process, culminating in the development of the multilaminar mammalian pallium. In all mammals studied so far, the cerebral cortex develops according to a common sequence (Boulder Committee 1969; reviewed by Caviness and Rakic 1978; Goffinet 1984b). As soon as migrating neurons reach the level of the cortical plate, they assume a remarkable radial differentiation (Shoukimas and Hinds 1978; Goffinet 1979). Marin Padilla (1978) first suggested that the developing mammalian pallium proceeds through a transient stage which he named the "primordial cortical organization". This transient stage would "recapitulate" the primitive cortex which is found in reptiles.

Comparisons between normal mice and animals affected by the reeler trait (a recessive mutation in mice which maps on chromosome 5) at critical stages of cortical embryogenesis point to the importance of cell-cell interactions for the organization of normal architectonic patterns (Goffinet 1979, 1980, 1984b). Briefly, the radial organization of cortical cells is defective in reeler mutants. This anomaly is seen as soon as cortical neurons end their radial migration through the early telencephalon. There is extensive evidence that the reeler anomaly does not result from abnormal cell generation, cell death or aberrant fiber growth and connections. An important event, somehow abnormal in reeler, must occur at the end of radial neuronal migration in order to "stabilize" early neural cell patterns. Although very little is known about this important morphogenetic step, the very fact that it is defective in reeler mutants suffices to show that it is submitted to a stringent genetic control (see Goffinet 1983, 1984a,b for discussion). As a corollary to this observation, the question of a possible phylogenetic control may be raised. In other terms: Is the property of early radial organization in the developing cortex present in every cortex, or has it been acquired during evolution of the cortex, independently of appearance of the cortex itself? Information on this question may be

Experimental Brain Research Series 19

gained by examining the development of the cerebral cortex in reptiles.

Although living reptiles do not bear any direct ancestral relationship to mammals, it is reasonable to assume that developmental features which are common to widely separate species might allow the definition of some general characteristics (a "common denominator") of corticogenesis. Furthermore, as will be clear after an overview of the main reptilian filiations (vide infra), similarities in cortical histogenesis between mammals and highly evolved reptiles such as lacertilians must result from evolutionary convergence and, as such, point to efficient solutions to some key problems arising during brain evolution (Goffinet 1983, 1984a; Goffinet et al. 1986a,b).

The most elegant way to examine this evolutionary question would be to compare brain development in mammals and in their reptilian ancestors. However, since mammalian ancestors are extinct, the developmental question formulated above can only be approached indirectly by comparing cortical histogenesis in the main reptilian phyla. Therefore, the results of our studies need to be understood by reference to known filiations among living reptilian groups.

Many points in the paleontology of reptilian evolution remain unknown and cannot be deduced from the observation of living species. For this reason, only well-established features will be considered for the present analysis. As a basis for discussion, we shall use the "minimal consensus" cladogram shown in Fig. 1, which was composed with the assistance of Prof. B. Cox. A common reptilian ancestor (probably during the Pennsylvanian) gave rise to several independent lines, four of which led to living reptiles, mammals and birds. A first branch, the synapsids, separated early and gave rise to mammals; a second branch led to chelonians, via a poorly understood lineage; a third branch led to Rhynchocephalia (of which *Sphenodon punctatus* is the only living remnant), as well as to lizards and their ophidian derivates; the fourth branch gave rise to the crocodilians (via thecodonts) and to birds (via saurischian dinosaurs). The third and fourth branches form the diapsid lineage, the ancestors of which correspond to eosuchians. The relationships between diapsids are still incompletely known. It is generally thought that chelonians (turtles) have evolved very little since the permian-triassic period and can be considered quite similar to stem reptiles, including mammalian ancestors. In order to cover the various branches of the evolutionary tree, we have examined cortical development in at least one species of turtles, squamates, crocodiles

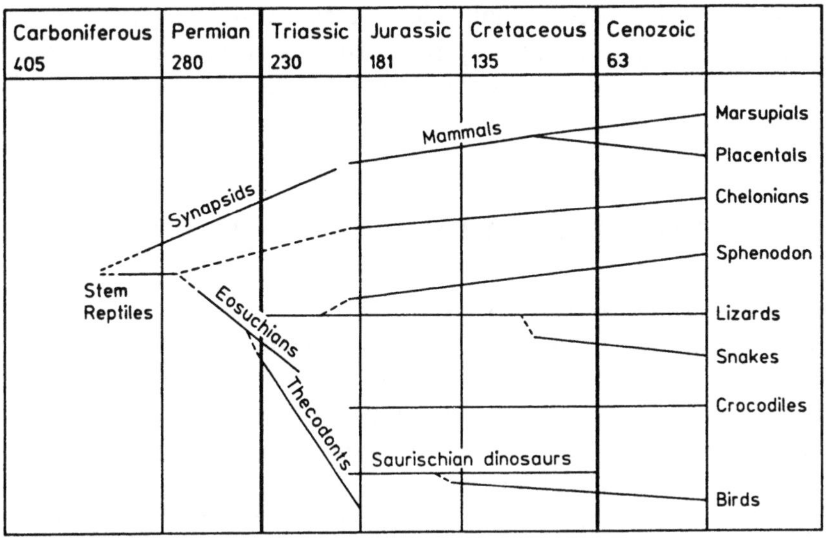

Fig. 1. Evolutionary filiations of upper vertebrates. Time in millions of years since beginning of geological period. See text for discussion.

and in *Sphenodon*. In addition, two representative species of the synapsid and squamate lineages have been analyzed in detail, using usual histological techniques, the Golgi method, electron microscopy and tritiated thymidine to determine the timing of neuronal histogenesis (Goffinet 1983, 1984a; Goffinet et al. 1986a,b).

Previous studies have dealt with the development of the cerebral cortex in various reptilian species. Telencephalic development has been examined in turtles by Kirsche (1972); in *Sphenodon punctatus* by Hines (1923), in the American alligator by Reese (1910), in several lizards by Faul (1926) and Hetzel (1974) and in the water snake by Warner (1946). A few interspecies comparisons have been attempted by Källen (1951) and Bergquist (1954). However, these works largely predate the elaboration of present views on mammalian cortical development. They are primarily focused on questions of neuroanatomical development and barely address issues such as cell differentiation and cortical histogenesis.

On the other hand, many morphological analyses of the cerebral cortex of adult reptiles have been carried out. They have yielded useful information on the connections and histology of the reptilian cortex and on homologies of telencephalic structures in reptiles, mammals and birds. This field, which is not directly relevant to our question, has been well reviewed by Northcutt (1981). Among other things, comparative anatomical data (Goldby and Gamble 1957) suggest that, among reptiles, turtles and lizards represent a rudimentary and a highly evolved type, respectively, of cortical architectonic organization.

Cortical Histogenesis in Reptiles

Embryological Observations

It is generally admitted that the cerebral cortex (or pallium) of reptiles is composed of four divisions: the dorsomedial cortex or hippocampus, the dorsal or general cortex, the lateral or pyriform cortex and the dorsal ventricular ridge (DVR) formerly called hyperstriatum. The DVR is included among cortical areas because it contains several sensory fields and is presumably phylogenetically related to pallial structures (for a discussion of the DVR, see Ulinski 1983).

For reasons mentioned above, cortical development was studied comparatively in several reptilian species belonging to the various branches of the evolutionary tree. Morphological studies at critical stages of embryonic cortical development show that important species variations are present in the degree of radial organization of the cortical plate (Fig. 2). The architectonics of the embryonic cortex is the most primitive in turtles. In these species, the radial cytological pattern is extremely poor, at the level of both neurons and glial cells. By contrast, all lizards have a nicely ordered cortical plate at all developmental stages. This property is best appreciated in Golgi preparations which show the radial deployment of neuronal processes as well as of radial glial fibers. The level of cortical organization in other reptilian species can be described as intermediate. For example, snakes are in this respect quite similar to lizards, whereas the crocodilian cortex is characterized by a poor architectonic pattern. In *Sphenodon,* the medial cortex is extremely well developed and organized, while the dorsal cortex is rather reminiscent of the primitive, turtle type.

Fig. 2a-d. The embryonic reptilian cortex: late embryonic stage. At a late embryonic stage, the turtle cortical plate (**a**,**c**) has a very primitive type of architectonic organization when compared with that in lizards (**b**,**d**). This difference is evident both at the level of the dorsal or general pallium (**a**,**b**) as well as the hippocampus (**c**,**d**). H&E stain. Magnification, x290

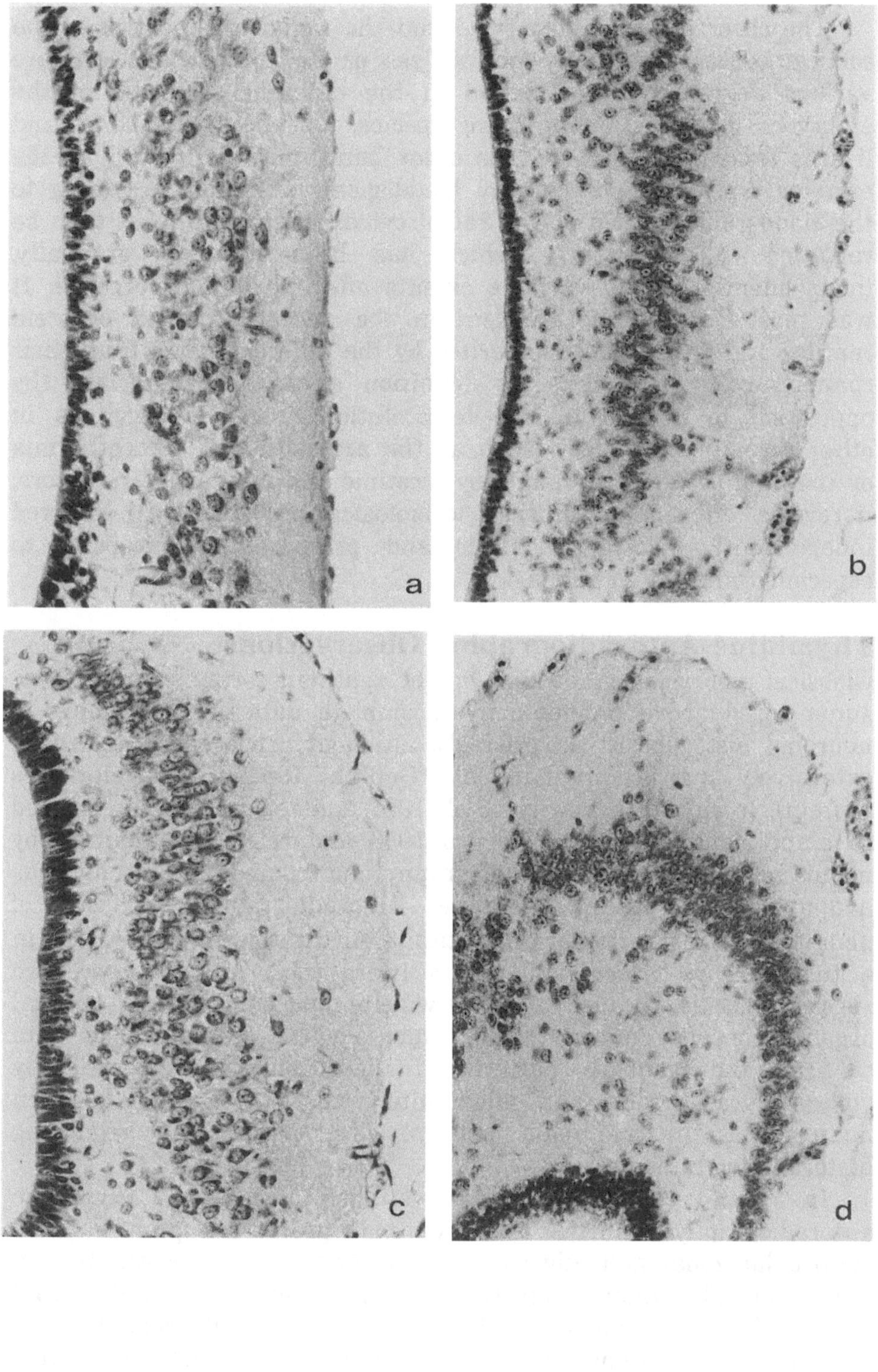
a
b
c
d

The observations demonstrate that the radial organization of the cortical plate is certainly not a "yes or no" property. There are various degrees of organization in the different segments of the embryonic pallium. Among living species, turtles, on one hand, and lizards (and mammals), on the other hand, may be viewed as the extreme types of a spectrum of histological variations. According to the cladogram outlined above, radial cytoarchitectonics must then be regarded as a property which has been acquired gradually, independently and to variable extents after phyletic divergence. It was most probably not present in the ancestral cortex of stem reptiles and has not been inherited by the different phyla from their common root. It is thus by definition a case of homoplasy (by opposition to homology), due to evolutionary convergence. As in other cases of homoplasy, such as the acquisition of homeothermia or the binding of acetylcholine by nicotinic and muscarinic receptors, it reveals that similar solutions to biological problems have occurred independently (Northcutt 1981) and presumably correspond to especially efficient solutions.

Thymidine Autoradiographic Observations

Classical embryological studies on the reptilian cortex such as those summarized above provide only incomplete data on the timing of neuronal histogenesis. Surprisingly, although neurogenesis has been extensively analyzed in mammals (Goffinet 1984a,b and references therein), in chick (Yurkewicz et al. 1981 and references therein) and in amphibians (Alley and Barnes 1983 and references therein) by means of thymidine autoradiography, no comparable study was available in reptiles. Therefore, autoradiographic analyses of neurogenesis in the forebrain cortical structures were carried out in a turtle (*Emys orbicularis*) and a lizard (*Lacerta trilineata*), two species selected because of their widely different types of cortical organization, as defined in the embryological analysis. The data confirm the general pattern of neurogenesis suggested by embryological studies and allow interesting comparisons between neurogenesis in mammals and reptiles, particularly regarding histogenetic gradients.

In both reptilian species studied, tritiated thymidine autoradiography confirms that neuronal precursors proliferate in ventricular zones and migrate radially towards their destination in the cortex. This radial migration most probably follows radial glial guides which are demonstrated in the embryonic telencephalic wall by electron microscopy as well as with the Golgi method. In this respect, radial migration may follow principles of guidance similar to those which are thought to act in the mammalian pallium.

With the relative exception of the medial cortex of lizards, where neurogenesis is prolonged for quite a long time, the bulk of neurogenesis proceeds within a time interval of less than 9 days (incubation temperature: 28°C). There is a relative exception to this general observation, in that neurogenesis in the medial cortex extends over a prolonged period. In embryos injected with thymidine before this stage of the "neuronal burst", all cortical neurons are labeled, indicating that only precursors are present in the ventricular zones at this time. Thymidine administration later than the stages of neurogenesis results only in the labeling of glial and mesodermal cells, so that ependymal, astrocytic and satellite cells differentiate or at least continue to proliferate after the majority of the neurons are formed. A striking observation was that in *Emys* embryos injected at late somite stages, an atrophy of the cerebral cortex was present at hatching. This might result from a radiotoxic effect of tritiated thymidine on small embryos, analogous to that described in chick embryos (Sauer and Walker 1961). The pattern of a "neurogenetic burst" is quite comparable to that seen in mammals. In mammalian cortical structures, only neurons of the dentate gyrus are generated over an extended time interval. This suggests that the dentate gyrus in mammals may be homologous to the medial cortex in reptiles, a theory which was already proposed by comparative anatomists (e.g., Goldby and Gamble 1957).

In the three supraventricular cortical fields (medial, dorsal and lateral cortices), neurogenesis follows a lateral to medial gradient and an anterior to posterior gradient (Fig. 3). These tangential gradients correlate with the pattern of development seen in embryological preparations of reptiles and mammals. In addition to the tangential gradients, radial histogenetic gradients are also found. In the lateral and dorsal cortices, radial histogenesis follows an outside to inside gradient. That is to say, late-generated cells settle at deeper levels in the cortex than older cells. This gradient is also perceived when synapse development is studied with the electron microscope. Synapses appear first in the molecular layer, then in the subcortex, and finally within the cortical ribbon. At the level of the medial cortex, however, a difference is apparent between the primitive cortical type (turtles) and the more elaborate pattern (lizards). In turtle cortex, an outside to inside gradient is evident. By contrast, it is impossible to define any gradient in the lacertilian medial cortex. The histogenetic gradients are best appreciated on charts of thymidine-labeled cells (Fig. 3). The mammalian cortex is

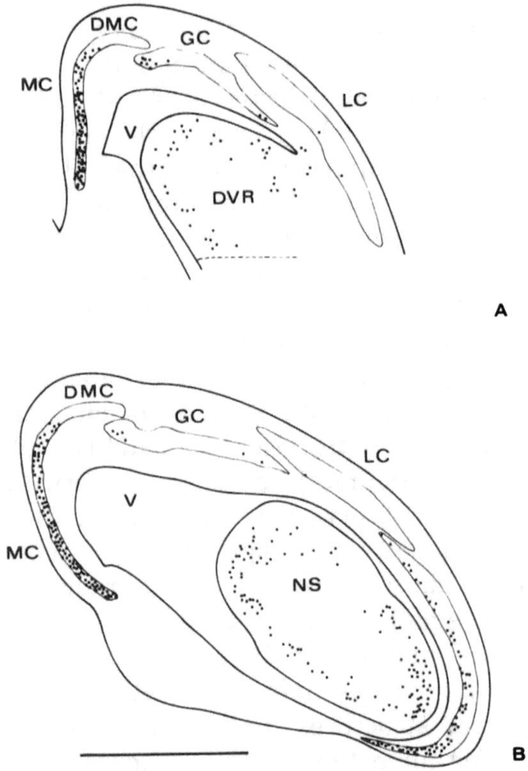

Fig. 3A,B. Example of camera lucida drawings with chart of labeled cells in frontal sections at anterior (**A**) and posterior (**B**) levels of a lizard specimen that received a thymidine pulse at stage 34 (end of neurogenesis). Labeled cells are indicated by *dots*. *MC*, medial cortex; *DMC*, dorsomedial cortex; *GC*, dorsal cortex (=general cortex); *LC*, lateral cortex; *DVR*, dorsal ventricular ridge; *NS*, nucleus sphericus; *V*, ventricle. *Bar* = 500 μm. Note the late development of the MC and NS.

characterized by the presence of an "inverted" histogenetic gradient (see Caviness and Rakic 1978 for discussion). That is, younger cells traverse layers of previously established neurons to settle at progressively more superficial levels. The present analysis demonstrates that this important property of the mammalian cortex is absent in reptile cortex, particularly in turtles which are the most closely related to stem reptiles and mammalian ancestors. If the phylogenetic assumptions outlined above are acceptable, it follows that the presence of an "inverted" histogenetic gradient must be considered as an acquisition, as a step in the evolution of the cerebral cortex from the stem reptiles to the mammals (synapsid radiation). Histological studies show that, among the divisions of the reptilian cortex, radial architectonics is the most elaborate in the lizard's medial pallium (Goffinet 1983a, 1984a). The autoradiographic analysis shows that the radial, outside to inside histogenetic gradient is absent in the same medial cortical area (Goffinet et al. 1986a,b). There is thus a gross negative correlation between radial laminar organization and the presence of an outside to inside gradient. It is worth reminding that an analogous situation is present in the mouse cortex. The normal mouse has a radially organized cortex which is formed following an inverted gradient. However, when radial cortical organization is disturbed, as in the reeler mutant mouse, histogenesis no longer follows the normal inverted gradient, but proceeds from outside to inside as it does in the most primitive cortices. The data thus strongly suggest that the evolutionary acquisition of an inside to outside pattern of cortical development is contingent upon, and largely determined by the radial organization of the embryonic cortical plate and of its constituent neurons (see Goffinet 1984b for discussion).

Conclusion and Perspective

As a tentative conclusion to these comparative studies of cortical development in reptiles as compared with mammals, we propose that radial organization, which so typical of the mammalian cortex, is not a general property of every cortex "per se", but rather the product of a progressive evolutionary process. As such, it must have its own genetic basis. Brain evolution occurs not only by increasing the number, the variety and the degree of differentiation of neural cells, but also by changing the geometry of neuron organization within the various brain nuclei. This process of evolutionary improvement of cell patterns, which probably results in a better efficacy of the brain, may be called "architectonic evolution". Further studies should aim at a better characterization of the biological factors involved in the determination of cortical cell patterns, some of which are defined operationally by our morphological analyses. The characterization of the reeler gene product would probably be useful in this respect.

References

Alley KE, Barnes MD (1983) Birthdates of trigeminal motoneurons and metamorphic reorganization of the jaw myoneural system in frog. J Comp Neurol 218:395-405

Bergquist H (1954) Notes on the early and morphogenesis of the central nervous system in vertebrates. J Comp Neurol 100:627-660
Boulder Committee (1969) Embryonic vertebrate central nervous system: revised terminology. Anat Rec 166:257-261
Caviness VS, Rakic P (1978) Mechanisms of cortical development: A view from mutations in mice. Annu Rev Neurosci 1:297-326
Faul J (1926) The comparative ontogenetic development of the corpus striatum in reptiles. Proc Acad Sci Amsterdam, Sect Sci 29:150-162
Goffinet AM (1979) An early developmental defect in the cerebral cortex of reeler mouse. Anat Embryol (Berl) 157:205-216
Goffinet AM (1980) The cerebral cortex of the reeler mouse embryo: an electron microscopic analysis. Anat Embryol (Berl) 159:199-210
Goffinet AM (1983) The embryonic development of the cortical plate in reptiles: a comparative analysis. J Comp Neurol 215:437-452
Goffinet AM (1984a) Phylogenetic determinants of radial organization in the cerebral cortex. Z Mikrosk Anat Forsch 98:909-925
Goffinet AM (1984b) Events governing the organization of postmigratory neurons. Brain Res Rev 7:261-296
Goffinet AM, Daumerie C (1986a) Genetic and phylogenetic determinants of radial cortical organization. In: Levi-Montalcini R, Calissano P, Kandel ER, Maggi A (eds) Molecular aspects of neurobiology. Springer, Berlin Heidelberg, New York pp 143-148
Goffinet AM, Daumerie C, Langerwerf B, Pieau C (1986b) H-3 thymidine autoradiographic analysis of neurogenesis in reptilian cortical structures. J Comp Neurol 243:106-116
Goldby F, Gamble HJ (1957) The reptilian cerebral hemispheres. Biol Rev 32:383-420
Hetzel W (1974) Die Ontogenese des Telencephalons bei *Lacerta sicula* (Rafinesque). Zool Beitr 20:361-458
Hines M (1923) The development of the telencephalon in *Sphenodon punctatus*. J Comp Neurol 35:483-535
Källen B (1951) On the ontogeny of the reptilian forebrain. Nuclear structures and ventricular sulci. J Comp Neurol 95:307-348
Kirsche W (1972) Die Entwicklung des Telencephalons der Reptilien und deren Beziehung zur Hirn-Bauplanlehre. Nova Acta Leopold 204:1-78
Marin Padilla M. (1978) Dual origin of the mammalian neocortex and evolution of the cortical plate. Anat Embryol (Berl) 152:109-126
Northcutt RG (1981) Evolution of the telencephalon in nonmammals. Ann Rev Neurosci 4:301-350
Reese AM (1910) Development of the brain of the American alligator. Smithson Misc Collect 54:3-20
Sauer FC, Walker BE (1961) Radiation injury from nuclear labeling with triated thymidine in the chick embryo. Radiat Res 14:633-642
Shoukimas GM, Hinds JW (1978) The development of the cerebral cortex in the embryonic mouse: an electron microscopic serial section analysis. J Comp Neurol 179:795-830
Ulinski PS (1983) Dorsal ventricular ridge: a treatise on forebrain organization in reptiles and birds. Wiley, New York
Warner FJ (1946) The development of the forebrain of the American water snake *Natrix sipedon*. J Comp Neurol 84:385-418
Yurkewicz L, Lauder JM, Marchi M, Giacobini E (1981) Triated thymidine long survival autoradiography as a method for dating the time of neuronal origin in the chick embryo: the locus coeruleus and cerebellar Purkinje cells. 203:257-267

Comparative Study on the Golgi Architecture of Avian Telencephalic Visual Areas

Teréz Tömböl
1st Department of Anatomy, Semmelweis University Medical School, H-1450 Budapest, Hungary

In the telencephalon of the avian brain, two visual centers are described: the ectostriatum and the visual Wulst. Separate visual pathways enter the telencephalon connecting the periphery with these visual areas: the retino-tecto-rotundo-ectostriatal pathway and the retino-thalamo- telencephalic tract. Numerous studies have been carried out on visual pathways since the first description of Ariens Kappers (1921), including electrophysiological (Revzin and Karten 1966/67; Engelage and Bischof 1988), lesioning and degeneration (Karten and Revzin 1966; Revzin 1969; Karten and Hodos 1970; Hunt and Webster 1972; Karten et al. 1973; Bagnioli and Burkhalter 1983; Macko and Hodos 1984) and tract tracing (Benowitz and Karten 1976; Nixdorf and Bischof 1982). Some of these studies involved behavioral experiments to determine the nature and extent of visual information processing (Hodos and Karten 1966, 1970; Hodos et al. 1988).

The ectostriatum is part of the non-laminar pallium (Hodos and Karten 1966, 1970; Karten and Hodos 1970; Hodos et al. 1973) and on the basis of its connections it would seem to be homologous with the thalamo-recipient layer IV of mammalian extrastriate cortex. In detailed analyses of the pathways, a central core (*ectostriatum centrale,* Ec) and a peripheral coat (*ectostriatum periphericum*, Ep) of the ectostriatum have been separated. The former is the recipient region of the tecto-rotundo-telencephalic pathway, while the latter receives fibers from the visual Wulst (Karten and Revzin 1966; Karten and Hodos 1970; Reiner and Karten 1983). In an analysis of Golgi preparations, a complex neuronal structure was found in the ectostriatum (Tömböl et al. 1988).

The visual Wulst is a projection area which is related to the visual cortex of mammals (Revzin 1969; Karten et al. 1973; Cuénod 1974; Pettigrew and Konishi 1976). To this area, the thalamo-fugal pathway ascends from the anterior dorsolateral nucleus of the

Experimental Brain Research Series 19

thalamus (Hodos et al. 1973; Pasternak and Hodos 1977). The Wulst is composed of several cytologically distinct laminae which have their individual names (Karten et al. 1973): *hyperstriatum accessorium, hyperstriatum dorsale,* and between them, the *hyperstriatum intercalare superius* and the *nucleus intercalatus hyperstriati accessorii.*

In the present study, an attempt is made to compare the Golgi structure of the two visual central areas. Special emphasis is put on the local neuronal network and on the GABAergic short-axon neurons. Finally, the avian and mammalian telencephalic visual centers are compared.

Material and Methods

For Golgi impregnation, 1-day and 28-day-old chickens were perfused with 4% formaldehyde transcardially. The brains were removed and impregnated in slices by rapid Golgi method. From the stained brain slices 100 μm thick serial sections were cut.

Four 1-day-old and two 14-day-old chickens were perfused with modified Hayat's fixative. Brains used for immunocytochemistry were kept in glutaraldehyde-free fixative overnight. After washes in phosphate buffer (4-6 h), 50-60 μm thick sections were cut with a vibratome. After suppressing the background staining, sections were incubated at 4°C with primary anti-GABA antiserum (GABA-g, raised and characterized by Hodgson et al. 1985) diluted 1:4000 in 1% normal goat serum for 24 h. Subsequently, sections were exposed to biotinylated goat anti-rabbit IgG diluted 1:200 in 1% normal goat serum for 2 h and then to avidin-biotin-peroxidase complex diluted 1:100 in 0.1 *M* phosphate buffer for 1 h. Immunoperoxidase was visualized by reaction with 0.05% 3-3'-diaminobenzidine in 0.05 *M* tris-HCl buffer containing 0.01% H_2O_2 for 5-10 min. After washing the sections were mounted on glass slides.

Figs. 1-7. Photomicrographs of neurons in Wulst and ectostriatum as demonstrated by Golgi impregnation (Figs. 1-6) and GAD immunostaining (Fig. 7). **1** Type 1a projection neuron in the Wulst. *Arrow* points to bifurcation of the axon. **2** Type 1b projection neuron in the Wulst. **3** Type 1c projection neuron in the Wulst. *Arrow* points to the axon. **4** Projection neuron in Ec. *Arrow* points to bifurcation of the axon. **5** Star-pyramid-like neuron in *hyperstriatum dorsale. Arrow* points to axon. **6** Large interneuron in ectostriatum. *Arrow*, axon bifurcation; *asterisks*, axonal boutons. **7** GABAergic neurons (*arrows*) and terminals (*asterisks*) in Ec

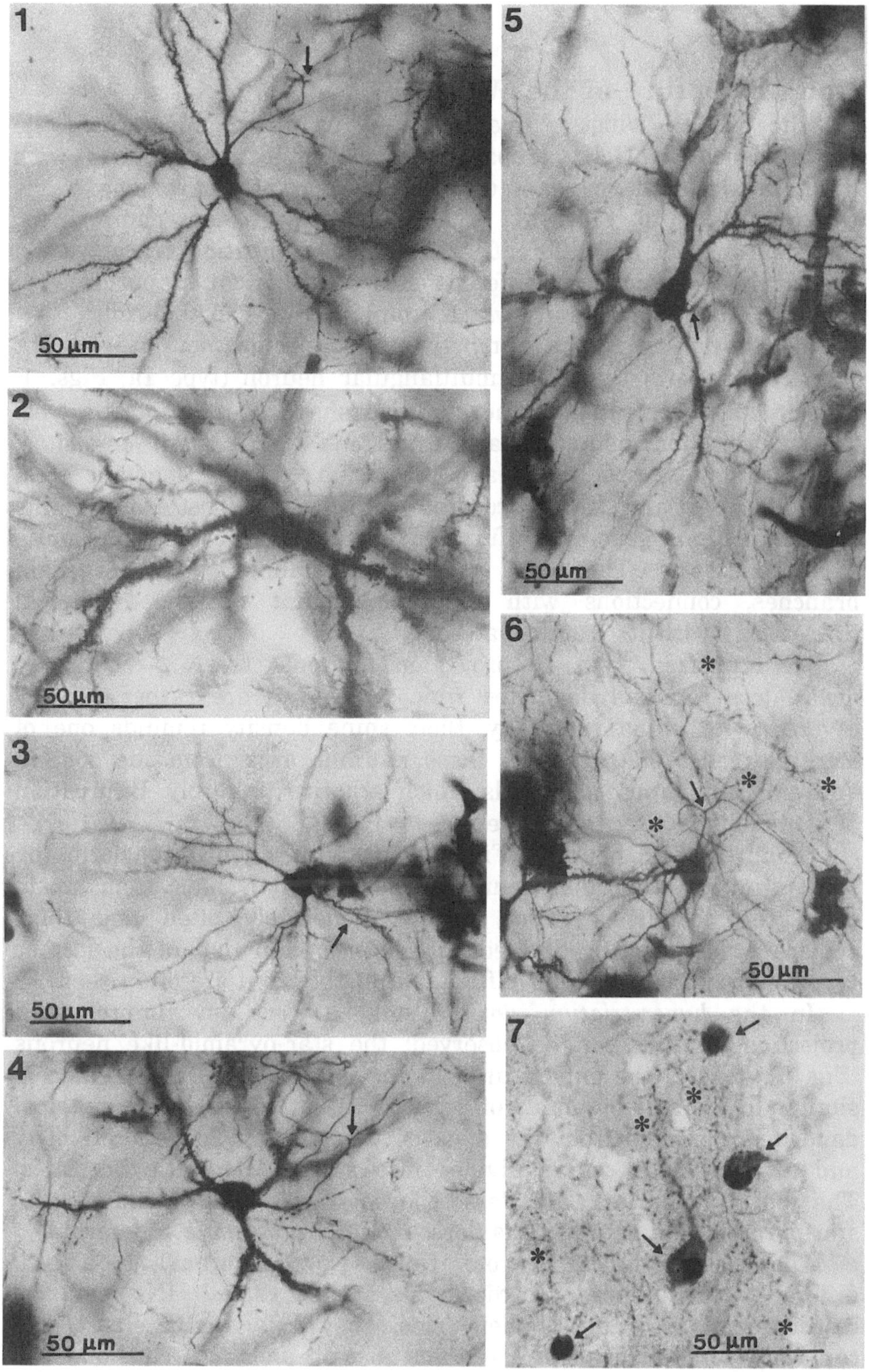
1
50 µm
2
50 µm
3
50 µm
4
50 µm
5
50 µm
6
50 µm
7
50 µm

Results

Golgi Structure of the Wulst

In the visual Wulst, three main types of neurons could be distinguished in Golgi preparations: (1) projection neurons with different size and shape of dendritic trees. They have long axons which project to different regions; (2) Short-axon neurons (or interneurons, INs) of different size of soma, dendritic trees and axon arborizations, which are different also in arborization pattern; (3) Small, stellate-like neurons which occur in groups in the Wulst.

(1) The most frequently impregnated projection neuron -- in our preparations -- is the large, multiangular neuron (type 1a, Figs. 1, 8), named so according to the size and shape of its dendritic tree. These projection neurons have numerous (8-10) main dendrites, which ramify several times and develop a complex dendritic tree. The dendrites are densely covered with spines. The axon of this type of neurons generally bifurcates after a short course and emits axon collaterals. The axon thus may establish, with its main branches, connections with different distant regions and local connections with its axon collaterals.

Medium-sized, type 1b projection neurons (Figs. 2, 9) display fewer, less ramified dendrites which, however, are rather densely covered with spines. Actually, their spine density reminds one of Purkinje cells. The axons of these neurons arise from the somata and emit numerous collaterals which arborize profusely. Bifurcation of the axons is very rarely seen.

Small, type 1c neurons (Figs. 3, 10), with a long axon form the third group of projection neurons. These neurons have from six to eight thin dendrites which ramify moderately often and their surface is less densely covered with spines than that of the former types. Axon collaterals and bifurcation of the axon are often seen.

In the *hyperstriatum dorsale* only a very few impregnated projection neurons can be observed: the star-pyramid-like neurons (Fig. 5). This is the only neuron in the bird telencephalon which is similar to the star-pyramids of layer IV of rat cortex. A short apical dendrite originates from the apex of the pyramid-shaped cell body and after a short course divides into smaller dendritic branches. The basal dendrites are less characteristic. The axon of the star-pyramid-like neuron runs dorsomedially and emits collaterals.

The INs (short-axon neurons; Golgi II type neurons) are also of different types: large, medium sized and small INs can be distinguished according to their sizes. They differ in the size of the cell body, in their dendritic tree and also in axon arborization. The

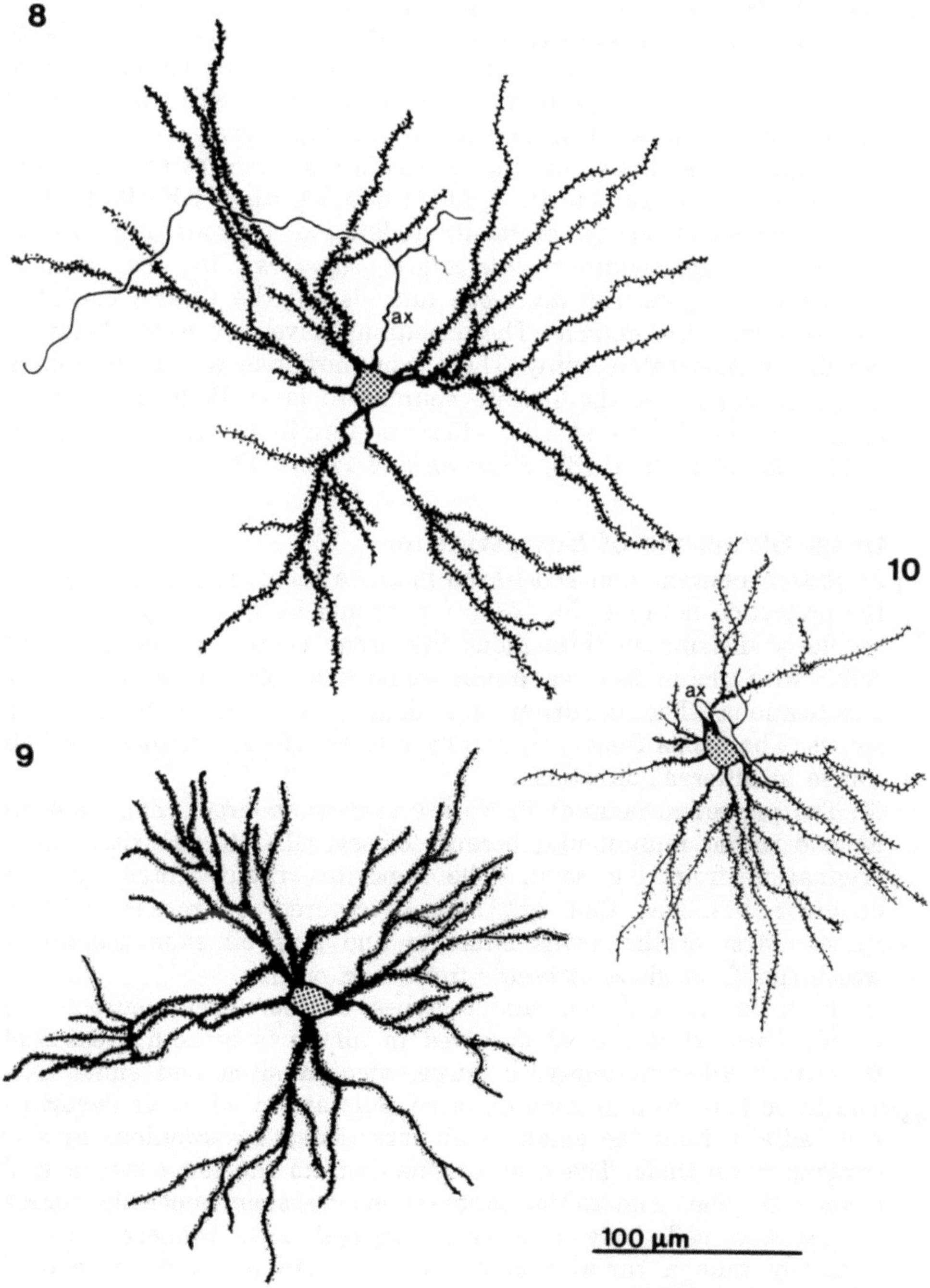

Figs. 8-10. Camera lucida drawings of Golgi impregnated projection neurons in Wulst. **8** Type 1a; **9** Type 1b; **10** Type 1c. *ax*, axons

smaller INs display a smaller axonal tree with intensely branched terminal segments, showing small, bulb-like varicosities. The large and medium-sized interneurons (Fig.11) have a larger axonal tree with long, varicose preterminal and terminal sections oriented rather parallel to each other. INs of this type have a similar axon arborization as those in the ectostriatum (the extent of axon arborization of large neurons is about 500-600 μm x 300-400 μm).

In the Wulst, groups of small, stellate-like neurons (Fig. 12) are found in *hyperstriatum intercalare superius*, in the *nucleus intercalatus hyperstriati accessorii* and also at the ventral border of the *hyperstriatum dorsale*. These neurons have few, wavy dendrites which are moderately spiny. Their axons arborize within the Wulst. They are similar to the stellate neurons in layer IV of mammalian cortex on which the specific afferents terminate (Ramón y Cajal 1911; Tömböl 1972, 1978b; Feldman and Peters 1978).

Golgi Structure of Ectostriatum

Projection neurons and short-axon neurons occur in this area. In Ec the projection neurons (Fig. 4) are morphologically homogeneous and regularly distributed throughout the area. They are medium sized cells with four to five main dendrites displaying only few ramifications. The dendrites are densely covered with irregular spines. The axons frequently ramify into equal-sized branches which course in different directions.

The projection neurons in Ep are somewhat larger than those in Ec and rather multangular because of several (7-8) dendritic shafts originating from the soma. The dendrites ramify intensely into dendritic branches that are densely covered by rather uniform spines. Most of the axons bifurcate and give off axon collaterals which ramify at short distances from their origins.

Both in Ec and Ep, similar types of short-axon neurons are found. These cells can be observed in all parts of both areas and they have different subtypes: large, medium-sized and small INs. The large (Fig. 6) and medium sized cells are ovoid, their dendrites run radially from the soma in all directions; few swellings appear irregularly on them. The axon originates from the soma with a well developed cone. The initial segment of the axon generally curves around the cell body and emits several side branches, which generally take a radial course and give rise to abundant axonal arborizations with further (seven to eight) bifurcations. The secondary and further axonal branches show numerous large swellings (3-4 μm long) and varicosities (1-2 μm long). These thickenings of the axon are most probably presynaptic sites. The

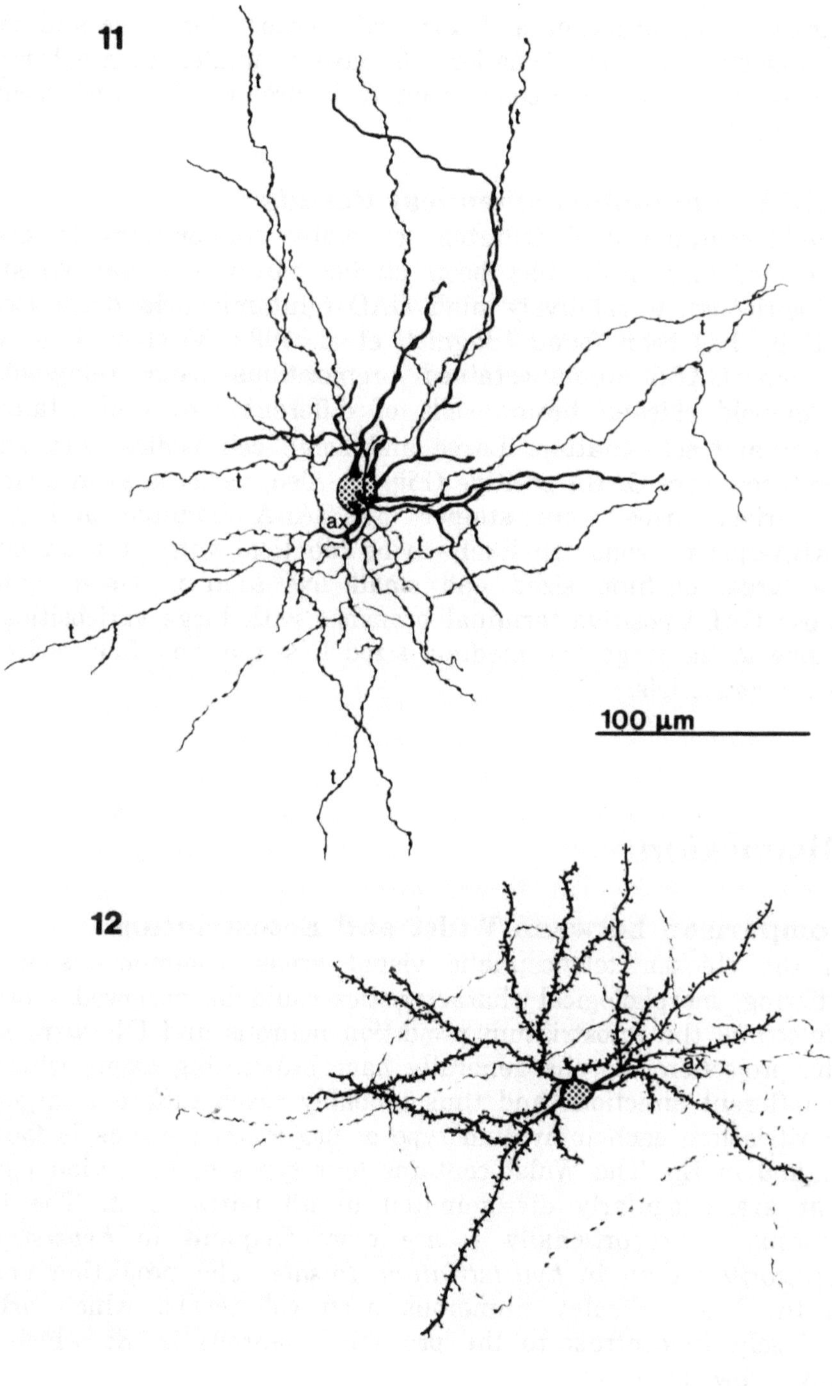

Figs. 11-12. Camera lucida drawings of Golgi impregnated interneuron **11** and stellate-like neuron **12** in Wulst. *ax*, axon; *t*, axonal boutons.

varicose axon branches and terminals contact dendrites and somata of projection neurons. Smaller INs have a smaller axon arborization domain and small varicosities on their preterminals and terminals (1-2 μm long).

GABA-Immunohistochemical Results

The topographical distribution of many transmitters in discrete areas of bird brain has been studied. In the visual Wulst and ectostriatum, a relatively high GAD (glutamic acid decarboxylase) activity has been found (Bagnioli et al. 1982; Vischer et al. 1982). In our GABA immunostained preparations from 1-day-old and 14-day-old chicken brains cells of different size were stained in Wulst and ectostriatum. Large and small cell bodies with smooth dendrites were GABA-positive (Fig. 7). Also, varicose axon terminals of various sizes were stained by GABA immunostaining. The GABAergic neurons are likely to be identical with at least part of the large, medium sized and small INs seen in Golgi material. Those GABA-positive terminal branches with large varicosities may belong to the large and medium sized INs, the very fine varicosities to the small INs.

Discussion

Comparison between Wulst and Ectostriatum

In the chicken telencephalic visual areas, common, as well as differing, morphological characteristics could be observed. Both the Wulst and the ectostriatum projection neurons and INs were found. The projection neurons generally have bifurcating axons which run in different directions and thus probably reach different targets. In ectostriatum, each individual type of projection neurons is found in Ec and in Ep. The Wulst contains four types of projection neurons that are irregularly disseminated in all parts of it. The larger neurons -- proportionally -- are more frequent in *hyperstriatum accessorium* than in *hyperstriatum dorsale*. The projection neurons in the Wulst display numerous axon collaterals, which arborize profusely, in contrast to the projection neurons in Ec which have only a few collaterals.

In both structures similar types of INs were found. The proportion of large INs is higher in ectostriatum than in Wulst. The large, medium-sized and small INs in both visual areas are probably GABAergic cells exerting inhibitory and/or modulatory effects, depending on the distribution of their axons. Thus, the large INs may suppress activity within a large zone in which the dendritic trees of several projection neurons overlap. Small INs have narrower fields of influence; they differ from the larger neurons in their axonal branching pattern and hence probably also in their functional relations. In any case, the complex IN distribution in the two visual areas points to a high degree of organization.

The morphology of projection neurons is more complicated in the Wulst than in the ectostriatum, the former containing three different types which probably establish local connections by their axon collaterals and distant connections by their long, bifurcating axons.

Comparison of Avian and Mammalian Visual Fields

The ectostriatum has been described as a functional equivalent of layer IV of peristriate mammalian cortex (Karten and Hodos 1970; Benowitz and Karten 1976; Watanabe et al. 1985). Layer IV of mammalian cortex contains stellate cells (Ramón y Cajal 1911) whose local axonal arborization is restricted to a vertical cortical column where connections with adjacent pyramidal and nonpyramidal neurons are established (Szentágothai 1975; Tömböl 1972, 1978a,b; Feldman and Peters 1978). On these stellate neurons, axo-spinous and axo-dendritic synapses are formed by the specific afferents (Feldman and Peters 1978).

In ectostriatum, there is no stellate-like neuron. The characteristic neurons in this area are the projection neurons. The visual pathway, the specific afferent of the ectostriatum, terminates on both, projection neurons and INs.

The visual Wulst has been homologized to the mammalian visual (striate) cortex (Revzin 1969; Karten et al. 1973; Cuénod 1974; Pettigrew and Konishi 1976) and the different Wulst areas (*hyperstriatum accessorium, hyperstriatum intercalare superius, hyperstriatum dorsale*) are believed, by some authors, to correspond to the cortical layers. The thalamofugal pathway in the Wulst has been suggested to terminate in a manner reflecting a functional "vertical-like" organization (Watanabe et al. 1983).

The layers of mammalian cortex consist of different types of neurons, but are generally dominated by one of these types, by the pyramidal neurons. A columnar organization of mammalian visual cortex can be observed in Golgi material and other histological preparations. The pyramidal neurons, with their apical dendrites, are the fundamental elements of the columnar construction. The ontogenetic development of their architecture is organized by radial glia cells which extend vertical processes up to the pia along which the neurons migrate (Rakic 1978; Tömböl 1980).

Comparing the Golgi structure of Wulst and mammalian cortex, one finds, at first sight, that there is no columnar arrangement of neurons in the Wulst and that pyramidal neurons are lacking. By contrast, multangular cells of different size, dendritic pattern and spine density are present. These projection neurons, like pyramidal neurons in mammalian cortex, display numerous, profusely arborizing axon collaterals for establishing local connections, whereas the main axon may bifurcate several times, providing the substrate for connections with different distant regions (association, commissural and subtelencephalic connections).

Although the Wulst, on the basis of Golgi data, seems to be very different from the mammalian visual cortex, it apparently contains morphological elements necessary for higher integrative activity, namely, the various types of projection neurons and INs which constitute the neuronal circuitry that receives and processes visual input.

References

Ariens Kappers CU (1921) Die vergleichende Anatomie des Nervensystems der Wirbeltiere und des Menschen. Bohn, Haarlem.

Bagnoli P, Burkhalter A (1983) Organization of the afferent projections to the Wulst in the pigeon. J Comp Neurol 214:103-113

Bagnoli P, Burkhalter A, Vischer A, Henke H, Cuénod M (1982) Effects of early monocular deprivation on choline acetyltransferase and glutamic acid decarboxylase in pigeon visual Wulst. Brain Res 247:289-302

Benowitz LJ, Karten HJ (1976) Organization of the tectofugal visual pathway in the pigeon. A retrograde transport study. J Comp Neurol 167:503-520

Cuénod M (1974) Commissural pathways in interhemispheric transfer of visual information in the pigeon. In: SchmittFO, Worder FG (eds) The neurosciences, third study program, MIT Press, Cambridge pp 21-29

Engelage J, Bischof H-J (1988) Enucleation enhances ipsilateral flash evoked responses in the ectostriatum of the zebra finch (*Taeniopygia guttata castanotis* Gould). Exp Brain Res 70:79-89

Feldman ML, Peters A (1978) The forms of non-pyramidal neurons in the visual cortex of the rat. J Comp Neurol 179:761-794

Hodgson AJ, Penke B, Erdei A, Chubb IW, Somogyi P (1985) Antiserum to γ-aminobutyric acid. I. Production and characterization using a new model system. Histochem Cytochem 33:229-239

Hodos W, Karten HJ (1966) Brightness and pattern discrimination deficits in the pigeon after lesions of *nucleus rotundus*. Exp Brain Res 2:151-167

Hodos W, Karten HJ (1970) Visual intensity and pattern discrimination deficits after lesions of ectostriatum in pigeons. J Comp Neurol 140:53-68

Hodos W, Karten HJ, Bonbright Jr. JC (1973) Visual intensity and pattern discrimination after lesions of the thalamofugal visual pathway in pigeons. J Comp Neurol 148:447-468

Hodos W, Weiss SRB, Besette BB (1988) Intensity difference thresholds after lesions of ectostriatum in pigeons. Behav Brain Res 30:43-53

Hunt SP, Webster KE (1972) Thalamo-hyperstriate interrelations in the pigeon. Brain Res 44:647

Karten HJ, Hodos W (1970) Telencephalic projections of the nucleus rotundus in the pigeon (*Columbia livia*). J Comp Neurol 140:35-52

Karten HJ, Hodos W, Nauta WJH, Revzin AM (1973) Neural connections of the "visual Wulst" of the avian telencephalon. Experimental studies in the pigeon (*Columbia livia*) and owl (*Speotyto cunicularia*). J Comp Neurol 150:253-278

Karten HJ, Revzin AM (1966) The afferent connections of the *nucleus rotundus* in the pigeon. Brain Res 2:368-377

Macko KA, Hodos W (1984) Near-field acuity after visual system lesions in pigeons. I. Thalamus. Behav Brain Res 13:15-30

Nixdorf BN, Bischof HJ (1982) Afferent connections of the ectostriatum and visual Wulst in the zebra finch (*Taeniopygia guttata castanotis* Gould): an HRP study. Brain Res 248:9-17

Pasternak T, Hodos W (1977) Intensity different thresholds after lesions of the visual Wulst in pigeons. J Comp Physiol Psychol 91:485-497

Pettigrew JD, Konishi M (1976) Neurons selective for orientation and binocular disparity in the visual Wulst of the barn owl (*Tyto alba*). Science 193:675-678

Rakic P (1978) Neuronal migration and contact guidance in the primate telencephalon. Postgrad Med J 54:25-40

Reiner A, Karten HJ (1983) The laminar source of afferent projections from the avian Wulst. Brain Res 275:349-354

Ramón y Cajal S (1911) Histologie du système nerveux de l'homme et des vertébres. Maloine, Paris

Revzin AM (1969) A specific visual projection area in the hyperstriatum of the pigeon. Brain Res 15:246-249

Revzin AM, Karten HJ (1966/67) Rostral projections of the optic tectum and the *nucleus rotundus* in the pigeon. Brain Res 3:264-276

Szentágothai J (1975) The "module-concept" in cerebral cortex architecture. Brain Res 95:475-496

Tömböl T (1972) A Golgi analysis of the sensory-motor cortex in the rabbit. In: Petsche H, Brazier MAB (eds) Synchronization of EEG activity in epilepsies. Springer, Vienna, New York, pp 25-36

Tömböl T (1978a) Comparative data on the Golgi architecture of interneurons of different cortical areas in cat and rabbit. In Petsche H, Brazier MAB (eds) Architectonics of the cerebral cortex, Raven Press, New York, pp 59-76

Tömböl T (1978b) Some Golgi data on visual cortex of the rhesus monkey. Acta Morph Acad Sci Hung 26:115-138

Tömböl T (1980) Some data on postnatal maturation of the cerebral cortex in cat. Acta Morph Acad Sci Hung 31:341-365

Tömböl T, Maglóczky Z, Stewart MG, Csillag A (1988) The structure of chicken ectostriatum. I. Golgi study. J Hirnforsch 29:525-546

Vischer A, Henke H, Cuénod M (1982) Neurotransmitter receptor ligand binding and enzyme regional distribution in the pigeon visual system. J Neurochem 38:1372-1383

Watanabe M, Ito H, Masi H (1983) Cytoarchitecture and visual receptive neurons in the Wulst of the Japanese quail (*Coturnix coturnix japonica*). J Comp Neurol 213:188-198

Watanabe M, Ito H, Ikushima M (1985) Cytoarchitecture and ultrastructure of the avian ectostriatum: afferent terminals from the dorsal telencephalon and some nuclei in the thalamus. J Comp Neurol 236:241-257

Differentiation and Selective Stabilization of Developing Brain Structures in Ducks. A Contribution to Ontogenetic Plasticity in the Avian Forebrain

G. Teuchert-Noodt, K.H. Breuker
and E. Kemming-Graebner
Dept. of Neuroanatomy, Faculty of Biology,
University of Bielefeld, D-4800 Bielefeld, FRG

Generally, the development of the vertebrate brain can be characterized as forming a sequence of numerous regressive and progressive events. When at about half-time of incubation the avian embryo becomes behaviorally active, regressive events (i.e. cell death, degeneration of synaptic elements) and progressive events (i.e. axonal sprouting, formation of new synapses, etc.) are equally involved in the development of brain systems (Rockstein and Sussman 1973; Jacobson 1978). This raises the question whether regression may play a part in the neuroplastic adaptation and stabilization of neuronal contacts.

Lysosome accumulation in degraded fibers and synapses is one regressive event which commonly appears during neuronal maturation of mammals (Wolff et al. 1989) and birds (Teuchert et al. 1983, 1986; Teuchert-Noodt 1989) and which enables visualization of synaptic dynamics. Applying a silver impregnation technique for the selective staining of lysosomes in degraded axon terminals, the occurrence of lysosome accumulation was monitored starting from the 3rd week of incubation up to the 4th week posthatching. Additionally, the differentiation and remodeling of neuronal connections was investigated in a rostro-medial subfield of the hippocampal complex.

The results are discussed in view of the question whether the regression of fibers and axon terminals and the survival of denervated neurons play parts in the adaptation and stabilization of neuronal contacts essential for brain function. Referring to recent experimental evidence for transient olfactory connections penetrating the rostro-medial hippocampal subfield (Teuchert et al. 1986), we

Experimental Brain Research Series 19

propose as the anatomical substrate of behavioral development a neuronal chain connecting sensory and association structures, and undergoing regression and progression processes.

Material and Methods

The topography of lysosome accumulation was studied in the brains of 33 embryos, 68 1-4-week old ducklings and seven 2-3 month-old ducks. Animals were perfused with 5% formaldehyde; brains were stored in the fixative for about 1-10 days and cut at a defined sagittal plane with a cryomicrotome (80 µm thick serial sections); sections were processed according to the silver impregnation method of Gallyas et al. (1980) and examined in the light microscope under dark-field illumination.

The extent of lysosome accumulation and cellular degradation in the hippocampal complex was analyzed in 19 embryos and ducklings by light and electron microscopy. After perfusion with 3% paraformaldehyde and 2.5% glutaraldehyde in cacodylate buffer (pH 7.2), brains were kept overnight in buffer. For later ultrathin sectioning, 1 mm thick parasagittal slices were taken from the tissue and embedded in Epon. In addition, semithin (0.5 µm) sections were cut and stained with p-phenylenediamine (Holländer and Vaaland 1968). For the study of cytological details of the hippocampal subfield slices from 28 brains of embryos from incubation days 26 to 28 and of newly hatched ducklings up to the age of 3 weeks were processed according to the Golgi-Kopsch method. Golgi impregnated cells were photographed and camera lucida drawn. Spines were counted on three 100-µm-long segments each of one apical and one basal dendrite in a total of 42 cells.

Results

Regional Lysosome Accumulation

Sensory projections During the 3rd and 4th week of incubation, lysosome accumulation in fibers and axon terminals progressed (1) from tectum to nucleus rotundus thalami, and from there via the thalamo-frontal tract to the ectostriatum (visual pathway; stippled in Fig. 1); (2) from nucleus mesencephalicus lateralis to striatal field L, and from there to hyperstriatum ventrale caudale (auditory pathway); (3) from the olfactory bulb to the prepiriform cortex and parolfactory lobe, which are the main olfactory centers of the avian forebrain (Pearson and Pearson 1976; Rieke and Wenzel 1978). From hatching onward, lysosome accumulations vanished from these structures except for a substantial number of lysosomes that persisted in and around the glomeruli of the olfactory bulb.

Hippocampal complex and forebrain laminae (dense dots in Fig. 1). From hatching onward, the pattern of lysosome accumulation proceeded beyond the olfactory forebrain fields and penetrated substructures of the hippocampal complex and both main forebrain laminae, i.e., lamina frontalis superior and lamina medullaris dorsalis (Teuchert et al. 1983; Teuchert et al.1986, 1989). Initially, distinct rows of accumulated lysosomes stood out in most medial planes of these structures, which together form part of the archicortex (Teuchert et al. 1985). In the somewhat older stages, the lysosome accumulations progressed to adjacent regions of the hyperstriatum, neostriatum and caudomedial (=limbic) archistriatum. These patterns vanished in ducklings about the age of 1 week.

Motor projections (broken lines in Fig. 1). About the age of 1 week, lysosome accumulations appeared in the anterior and intermediate motor regions of the archistriatum, and in the paleostriatal complex and lateral reticular formation from mesencephalic to medullary levels. An additional, intense staining pattern was found in both main motor projections passing through the anterior brainstem (i.e., the non-hypothalamic, occipitomesencephalic tract and the septomesencephalic tract). Some of these regions appeared homogeneously occupied by lysosomes for at least 2-3 weeks. The pattern disappeared at the age of about 4-6 weeks.

Evaluation

It is most conspicuous that in all sensory projections, patterns of lysosome accumulation arose at about half-time of incubation and slowed down markedly around hatching. In the motor projection, accumulation patterns appeared at later, posthatching stages. This difference may show that neuronal connections of afferent projections become stabilized first and neuronal connections of efferent projections become stabilized later. The adaptive value of the sequential occurrence of lysosome accumulations from embryos to young birds has to be seen in the precociousness of this avian species and the adaptation of these patterns to other developmental steps: While sensory systems must be perfectly functioning at hatching, motor abilities necessarily demand some kind of

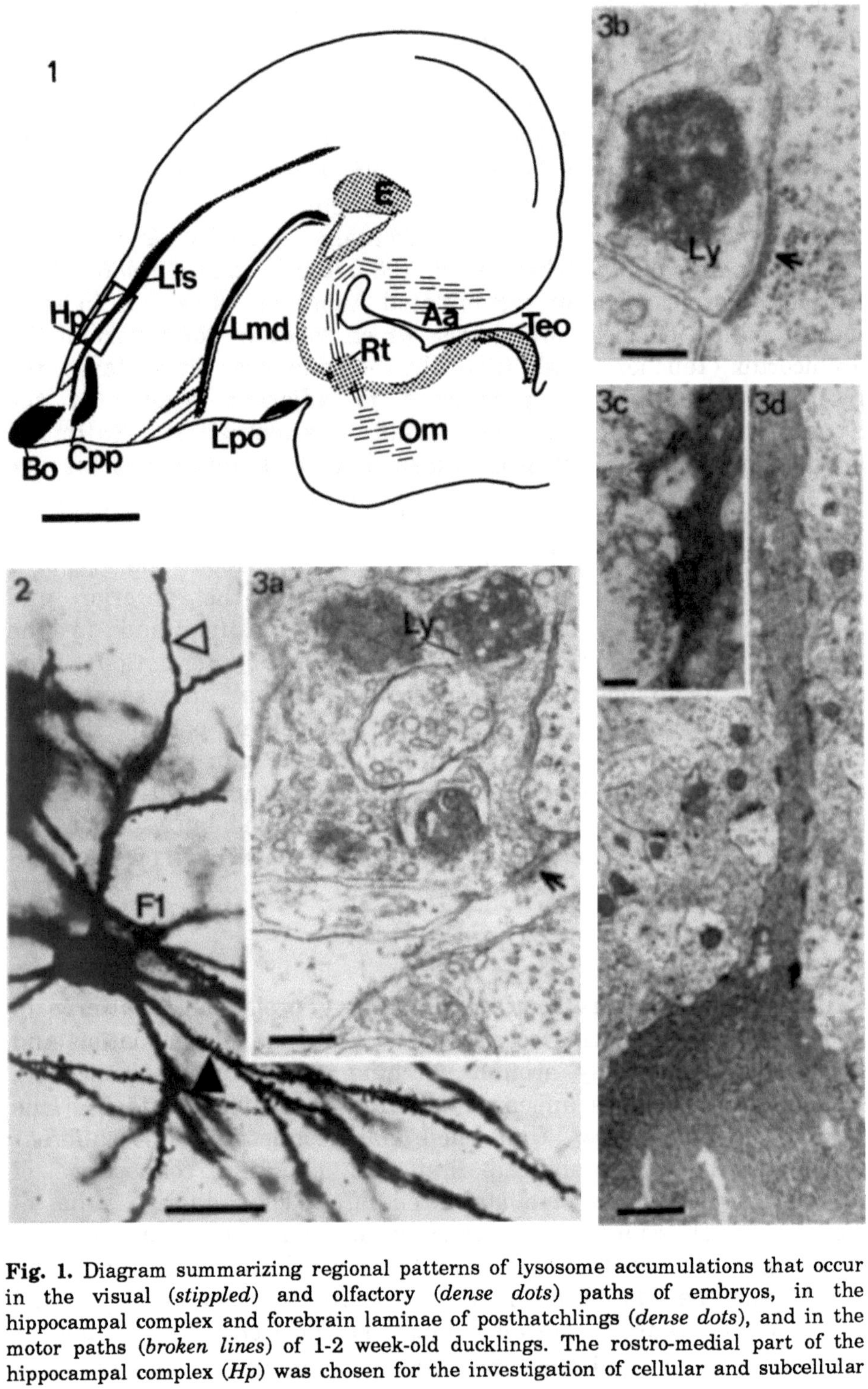

Fig. 1. Diagram summarizing regional patterns of lysosome accumulations that occur in the visual (*stippled*) and olfactory (*dense dots*) paths of embryos, in the hippocampal complex and forebrain laminae of posthatchlings (*dense dots*), and in the motor paths (*broken lines*) of 1-2 week-old ducklings. The rostro-medial part of the hippocampal complex (*Hp*) was chosen for the investigation of cellular and subcellular

conditioning in the young bird, resulting in a stepwise perfecting of its motor functions (ethological data in: Bekoff 1976,1982; Oppenheim 1974; Gottlieb 1979). The sequence of lysosome accumulation patterns thus offers a hint to the ontogenetic neuroplasticity which underlies the acquisition of integration in sensory motor function.

That lysosomes accumulate in the hippocampal complex and forebrain laminae just in between areas of accumulation in sensory and motor systems suggests that the former forebrain regions form a link between sensory and motor systems with regard to the adaptation of sensory motor abilities to high-order functions. The significance of the avian hippocampal complex in this developmental context has led to a more detailed examination of a small substructure of the rostro-medial forebrain (Teuchert-Noodt et al. 1989). Some results have been reported.

Differentiation of a Rostro-Medial Subfield of the Hippocampal Complex

The avian hippocampal complex is one of the telencephalic components about which we are still largely ignorant with regard to their structural and functional aspects (Ariens-Kappers et al. 1967; Pearson and Pearson 1976). Basically, the paleostriatum and limbic archistriatum are connected with this region in some way (Crosby and Humphrey 1939; Zeier and Karten 1971; Karten and Dubbeldam 1973; Benowitz and Karten 1976), and in diverse avian species this region is described as being highly adaptive to the molding of the olfactory sense (Craigie 1940). Moreover, olfactory

◄———

degradation as demonstrated in Figs. 2 and 3. *Aa*, archistriatum; *Bo*, bulbus olfactorius; *Cpp*, area prepiriformis; *E*, ectostriatum; *Lfs*, lamina frontalis superior; *Lmd*, lamina medullaris dorsalis; *Lpo*, lobus parolfactorius; *Om*, tractus occipitomesencephalicus; *Rt*, nucleus rotundus; *Teo*, tectum opticum. *Bar*, 0.5 mm

Fig. 2. Golgi staining of type-F cell (*F1*, fusiform cell) in the hippocampus of duckling. The cell has lost spines predominantly on somata and on several dendritic segments (*white arrow*); other segments maintain spine density (*black arrow*). *Bar*, 15 μm

Fig. 3a-d. a,b Electron microscopy of degraded axon terminals containing lysosome accumulations (*Ly*); note the preservation of postsynaptic thickenings (*arrows*). **c,d** Dark staining of type-F cells with synaptic contacts. *Bars*: **a** = 80 nm; **b** = 360 nm; **c** = 130 nm; **d** = 500 nm representation is described as mostly regressive in this region (Huber and Crosby 1929).

In experiments with duck embryos, the hippocampal complex was shown to sustain transient connections with the olfactory bulb via the medial olfactory tract, which becomes subject to massive degeneration during the designated developmental stage (Teuchert et al. 1986). On the basis of these results we suggest that the concomitant appearance of lysosome accumulation in the hippocampus corresponds to the natural degeneration of embryonic olfactory projections.

1. Light and electron microscopy. An overall cellular degradation was found in semi-thin sections stained according to the method of Holländer and Vaaland (1968). In the light microscope a large proportion of the affected cells appeared as dark neurons during the 1st and 2nd weeks after hatching. The persistent dark staining of these cells implied degradation with recovery rather than cell death. Electron microscopic examination (Fig. 3c,d) revealed various degrees from weak to strong degradation, i.e., electron density of somata caused by considerable concentrations of free ribosomes. Degradation of cellular organelles was accompanied by the occurrence of lysosomes in axon terminals and fibers. Both axo-somatic (Fig. 3b) and axo-dendritic (Fig. 3a) synapses were found to be affected; however, postsynaptic thickenings of the degraded terminals appeared unchanged. It may be emphasized that not all synaptic transformations which were indicated by the study of silver impregnation samples necessarily affected the entire terminal. Apparently, a certain proportion of spherical vesicles are preserved in the terminals. This supports the view of a partial recovery of degraded neurons.

2. Golgi staining. Starting with investigations of late embryonic stages it became obvious that hippocampal cells had developed the characteristics of pyramidal and fusiform cells, respectively, and had spines on somata and dendrites a few days before hatching. Dendritic trees of both types of neurons were preferentially located in the molecular layer and the intrinsic fiber layer of lamina frontalis superior. Dendritic spines of these cells may thus provide possible targets for axon terminals of embryonic olfactory projections.

From hatching onward, pyramidal and fusiform cells were found to lose spines predominantly on their somata and on several of their dendritic segments, whereas other segments maintained their spine density completely (Fig. 2). From about day 10 onwards, a lot of spines became detectable, and typical features of dendritic arborizations were retained. No recovery of spines, however, was observed on the somata of these cells. It seems thus reasonable to assume that hippocampal cells are subject to a reorganization of dendritic spines.

Discussion of a Model

From previous discussions of neuroplasticity in brain development (e.g., Cotman 1985) it has become increasingly likely that regression is an attendant circumstance of neuronal maturation and that - unlike ontogenetic cell death - subcellular regression plays some role in directly stimulating synaptogenesis (Wolff 1981). While a restricted proportion of fibers and terminals may disappear, the ability to compete for other, alternative, terminals will increase, thus providing the basis for a remodeling of neuronal connections (Dammasch et al. 1986; Wolff and Wagner 1983).

Our results on restricted patterns of lysosome accumulations in sensory motor projections suggest that the so-called "compensation theory of synaptogenesis" (Wolff and Wagner 1983) holds true for the adaptation of whole avian brain systems: a synaptogenic imbalance is produced by the lysosome degradation of axon terminals in sensory substructures first and in motor substructures later. This means that sensory systems become functional in duck embryos and motor systems in the young duckling. Furthermore, the spatio-temporal occurrence of the lysosome accumulation patterns implies that also other brain systems provide sensory motor integrations to a high-order level. This requirement is reflected by the occurrence of lysosome accumulation patterns in the hippocampal complex and forebrain laminae.

On the basis of the data obtained from our studies on the rostro-medial part of the hippocampal complex (Teuchert et al. 1986; Teuchert 1989; Wolff et al. 1989) we offer a neuronal chain model as the underlying principle of adaptation and stabilization of microcircuits in structural brain development. The model (Fig. 4) shows three cells in secondary sensory regions (S), which may represent specific sensory qualities (e.g., visual, auditory, somatosensory) and their neural paths. These cells are interconnected with three subordinate associative cells (A) in

4

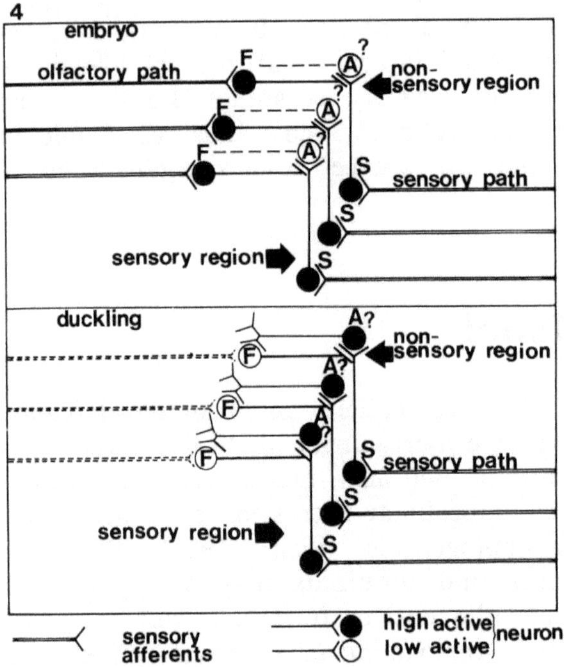

Fig. 4. Sensory pathways (*double lines*) and cells (*S*) having connections with associative cells (*A*) in nonsensory regions. Hippocampal cells (*F*) and olfactory paths (*double lines*) are involved in circuit remodeling: In the embryo (*top*) transient olfactory paths (*double lines*) exert a direct influence on type-F cells, and -- hypothetically -- a transneuronal inhibitory influence on type-A cells (*question mark*). In the duckling (*below*) olfactory paths are diminished (*interrupted double lines*) and the denervated type-F cells, now standing under low activity, are reinnervated by axonal growth of type-A cells.

nonsensory regions. Additionally, three hippocampal cells (F) are involved in the remodeling of neural connections.

In the embryo (Fig. 4, top) three transient olfactory paths to type-F cells are assumed. The existence of such projections has been demonstrated by Teuchert et al. (1986). The type-F cells possibly exert an inhibitory influence on type-A cells. In the duckling (lower half of Fig. 4), changes in connectivity become immediately effective when a proportion of transient olfactory paths is diminished. This has been demonstrated to occur around the time of hatching. At the same developmental stage, type-F cells have been found to become degraded and to lose a certain proportion of somatic and dendritic spines. The ability of type-F cells to compete for other terminals will increase as the available terminal space increases. The vigor of axonal growth of type-A cells which are influenced by the highly active input of the specific sensory paths will result in an increase of terminal arborization. This has been proven by the recovery of type-F cells from about day 10 onward.

The process of remodeling hippocampal connectivity may be subject to environmental influences. Reorganization of microcircuits could thus be an adaptational step in the development of higher-order structures and the integration of sensory-motor functions.

References

Ariens-Kappers CU, Huber GC, Crosby EC (1967) The comparative anatomy of the nervous system of vertebrates, including man. Hafner, New York, pp 1358-1401

Bekoff A (1976) Ontogeny of leg motor output in the chick embryo: a neural analysis. Brain Res 106:271-291

Bekoff A (1982) Verhaltensembryologie bei Vögeln und Säugern: Neuroembryologische Untersuchungen zur Entwicklung der Verhaltensmotorik. In: Immelmann K, Barlow G, Petrinovich L, Main M (eds) Verhaltensentwicklung bei Mensch und Tier, Das Bielefeld-Projekt. Parey, Berlin, pp 271-281

Benowitz LI, Karten HJ (1976) The tractus infundibuli and other afferents to the parahippocampal region of the pigeon. Brain Res 102:174

Cotman CW (1985) Synaptic plasticity. Guilford, New York

Craigie EH (1940) The cerebral cortex in paleognathine and neognathine birds. J Comp Neurol 73:179-234

Crosby EC, Humphrey T (1939) Studies on vertebrate telencephalon. J Comp Neurol 71:121-213

Damasch IE, Wagner GP, Wolff JR (1986) Self-stabilization of neuronal networks. Biol Cybern 54:211-222

Gallyas F, Wolff JR, Böttcher M, Záborsky L (1980) A reliable and sensitive method to localize terminal degeneration and lysosomes in the central nervous system. Stain Technol 55:299-306

Gottlieb G (1979) Development of species identification in ducklings. V. Perceptual differentiation in the embryo. J Comp Physiol Psychol 93:831-854

Holländer H, Vaaland JL (1968) A reliable staining method for semi-thin sections in experimental neuroanatomy. Brain Res 10:120-126

Huber GC, Crosby EC (1929) The nuclei and fiber paths of the avian diencephalon, with consideration of telencephalic and certain mesencephalic structures. J Comp Neurol 48:1-225

Jacobson M (1978) Developmental neurobiology, 2nd edn. Plenum, New York

Karten HJ, Dubbeldam JL (1973) The organization and projection of the paleostriatal complex in the pigeon (*Columbia livia*). J Comp Neurol 148:61-90

Oppenheim RW (1974) The ontogeny of behavior in the chick embryo. In: Lehrman DS, Hinde RA, Shaw E, Rosenblatt J (eds) Advances in the study of behavior, vol. 5. Academic Press, New York, pp 133-172

Pearson R, Pearson L (1976) The Vertebrate Brain. Academic Press, London

Rieke GK, Wenzel BM (1978) Forebrain projections of the pigeon olfactory bulb. J Morphol 158:41-56

Rockstein ME, Sussman ML (1973) Development and aging in the nervous system. Academic Press, New York

Teuchert G, Wolff JR, Immelmann K (1983) Physiologische Degeneration in der Ontogenie des ZNS von Vögeln - eine Einflußnahme auf die sensible Phase für Prägung. Verh Dtsch Zool Ges, 76:259

Teuchert G, Reissmann T, Vockel A (1985) Developmental aspects of the telencephalic laminae in birds. Fortschr Zool 30:589-591

Teuchert G, Reissmann T, Vockel A (1986) Olfaction in peking ducks *(Anas platyrhynchos)*: a comparative study of centrifugal and centripetal olfactory connections in young ducks and in embryos and ducklings (Aves). Zoomorphologie 106:185-198

Teuchert-Noodt G (1989) Central nervous system ontogeny. An evaluation of the recapitulation hypothesis. In: Duncker H-R, Fleischer G (eds) Functional morphology in vertebrates. Proc 2 Int Symp Vertebr Morphol, Vienna, 1986

Teuchert-Noodt G, Wolff JR, Breuker KH (1989) Sequential aspects of synaptic remodeling in the developing brain of ducks *(Anas platyrhynchos)* (in press)

Wolff JR (1981) Some morphogenetic aspects of the development of the central nervous system. In: Immelmann K, Barlow GW, Main M, Petrinovich L (eds) The Bielefeld interdisciplinary project. Behav Dev, (special issue) pp 164-190

Wolff JR, Wagner G (1983) Selforganization in synaptogenesis: interaction between the formation of excitatory and inhibitory synapses. In: Basar E, Flohr H, Haken H, Mandell AJ (eds) Synergetics of brain. Proceedings of the International Symposium on Synergetics at Schloß Elmau, Bavaria. Springer Berlin, Heidelberg, New York 50-59 (Springer series in synergetics, vol 23)

Wolff JR, Leutgeb M, Holzgraefe M, Teuchert-Noodt G (1989) Synaptic remodelling during primary and reactive synaptogenesis. Fortschr Zool 37 (in press)

Zeier H, Karten HJ (1971) The archistriatum of the pigeon. Organization of afferent and efferent connections. Brain Res 31:313-326

Basic Mechanisms of GABAergic and Cholinergic Influences on Sensory Telencephalic Processing in Chicken and Cat

Christian M. Müller
Max-Planck-Institut für Hirnforschung
Deutschordenstraße 46, D-6000 Frankfurt/M. 71, FRG

Experimental approaches towards an understanding of information processing in the CNS can follow two main routes: the study of one species or neuronal substrate specialized for a certain task or the comparison of analogue neuronal substrates in unrelated species with comparable information processing capacity. The latter approach has the advantage that similarities found in the way central nervous computation is performed in unrelated species may uncover **basic** principles of information processing in the brain. In the present report I will summarize recent findings on the action of two transmitters, gamma-aminobutyric acid (GABA) and acetylcholine (ACh), in the caudal auditory telencephalon of a bird and the visual cortex of a mammal.

The study of sensory systems has the advantage that the capacity of an experimental animal for sensory processing is accessible to behavioral studies. In addition, physiological experiments are facilitated by the ability to test neuronal responses with well-controlled stimuli, including those stimuli with behavioral relevance. Studies utilizing behavioral testing regimes have shown that telencephalic structures play a key role in the recognition of multiple behaviorally relevant sensory stimuli. Telencephalic lesions interfere with the capability to recognize complex visual stimuli (Sprague et al. 1977; Meadows 1974), species-specific communicational sounds (Dewson et al. 1969; Neff 1977), and localization of sound sources (Jenkins and Masterton 1982). This effect is consistent with the presence of neurons coding such behaviorally relevant sensory signals especially (for most signals even exclusively) in telencephalic structures (Scheich et al. 1979; Winter and Funkenstein 1973; Perret et al. 1982; Suga et al. 1983). These findings suggest that telencephalic processing is essential for recognition processes.

Experimental Brain Research Series 19

To understand the underlying central nervous processes in the telencephalon I studied the actions of two transmitter systems known to be involved in signal analysis. GABAergic inhibition has been shown to be essential for a variety of response specificities in telencephalic neurons. Application of GABA-receptor antagonists selectively abolishes neuronal selectivity for certain features of sensory stimuli characteristic of telencephalic neurons (e.g., Sillito 1977, 1979; Dykes et al. 1984). Thus, it is assumed that GABAergic inhibition is one causal mechanism of central nervous feature extraction. The cholinergic system is involved in recognition processes by mediating attention and arousal (Phillis 1968; Sato et al. 1987; Singer 1979). Interference with cholinergic transmission leads to sensory discrimination deficits in mammals (Bennett et al. 1971; Evans 1975) as well as in birds (Cleeves and Green 1982). The underlying mechanisms are, as yet, only poorly understood.

The studies summarized in the present report were conducted to further elucidate the mechanisms by which GABA and acetylcholine act in sensory processing, comparing findings in a nonmammalian vertebrate (chicken) and a mammalian vertebrate (cat). The similar effects of the transmitters and the integration of the transmitter actions in the functional organization of the telencephalic structures are discussed in terms of the basic roles GABA and acetylcholine may have in sensory processing in vertebrates.

Functional Organization of the Avian Caudal Auditory Telencephalon and the Mammalian Sensory Cortex

Studies were done in the caudal auditory telencephalon (CAT) of the chicken and the primary visual cortex (area 17) of the cat. As summarized in Fig. 1, these structures have several common features in their circuitry and functional organization. Both areas contain defined sensory representations; certain features of the sensory modality coded in these areas are mapped onto the extent of the structure. In the avian auditory forebrain there is a *tonotopic organization* (an orderly representation of stimulus frequencies extending from dorso-lateral to medio-ventral; see Bonke et al. 1979b; Müller and Leppelsack 1985) and a representation of *ear dominance* (Scheich 1983). In the mammalian visual cortex we observe an orderly representation of the visual space *(retinotopy)*, of *ocular dominance*, and of *orientation selectivity* (for review, see

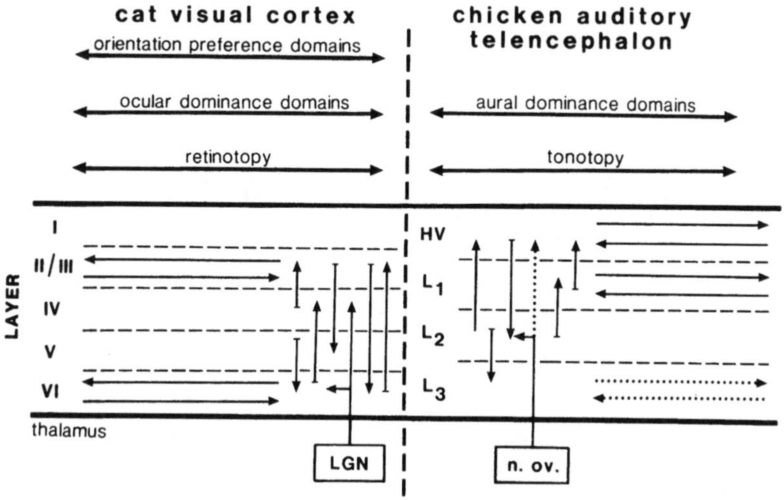

Fig. 1. Thalamo-telencephalic and intratelencephalic interconnectivity in the caudal auditory telencephalon of birds (based on: Bonke et al. 1979a; Müller 1987b) and the visual cortex of the cat (Gilbert 1983) share several features. In both species thalamic afferents terminate in selective layers of the laminated telencephalic structures. Layers (cat, I-VI; chicken, HV L1-L3) are interconnected by vertical projections, while reciprocal horizontal connections are present within specific laminae. Sensory representations of stimulus features, coded by single neurons, extend orthogonally to the layering of the structures. *LGN*, lateral geniculate nucleus; *n.ov.*, nucleus ovoidalis; *HV*, hyperstriatum ventrale

Hubel and Wiesel 1977). In both areas these functional topographies extend orthogonally to the layering of the structures. Four layers exist in the CAT of birds, three in the neostriatum and one in the adjacent hyperstriatum ventrale (see Fig. 2). The central neostriatal lamina (L2) receives the input from the thalamic nucleus ovoidalis (Bonke et al. 1979a). This layer thus corresponds to layer IV of the mammalian cortex, which is the main thalamo-recipient cortical layer. Efferents from layer L2 in the bird auditory forebrain, as well as layer IV of the mammalian cortex, innervate the adjacent layers. Intercolumnar projections *(vertical interconnections)* couple homotopic points in the functional topography of the cortex, as well as in the avian CAT. In addition, there are *horizontal interconnections* along the extent of the layers. These latter projections are usually reciprocally organized (Luhmann et al. 1986).

Thus, both areas under study are characterized by the presence of functional topographies orthogonal to a layering, and the presence of a similar set of both interlaminar (vertical) and intralaminar (horizontal) connections.

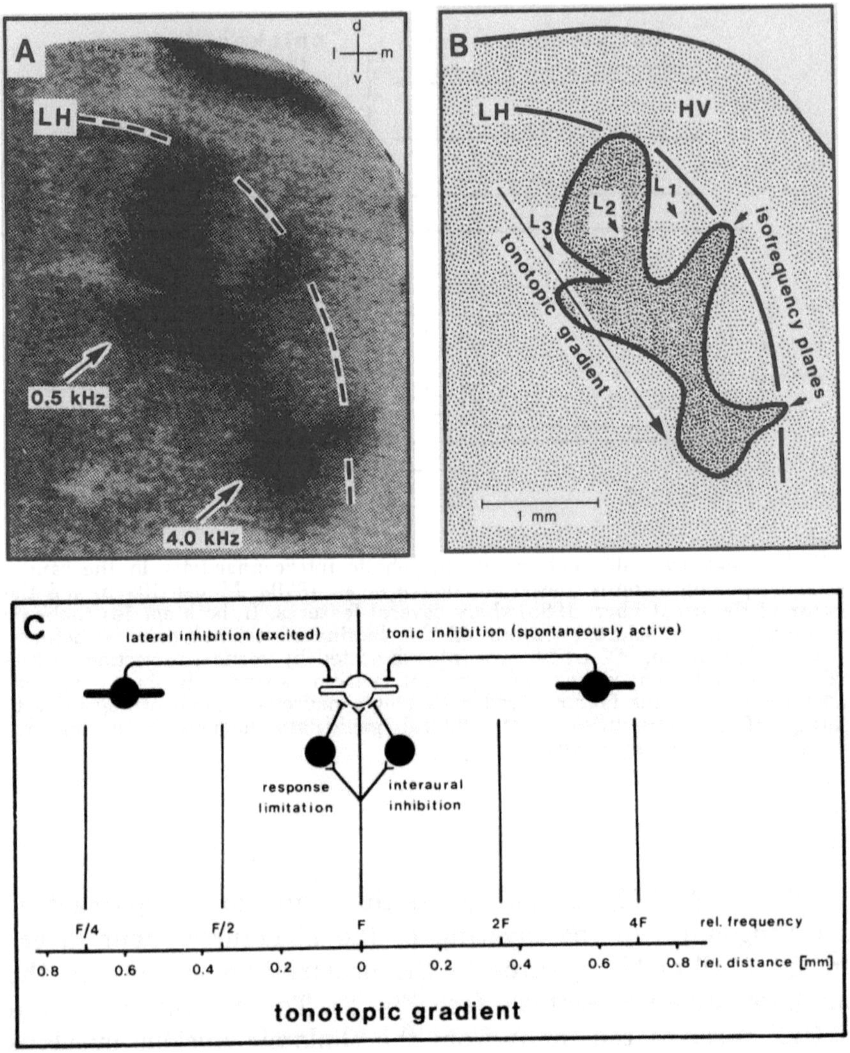

Fig. 2 A-C. Functional organization of the caudal auditory telencephalon (CAT) of the chicken, and integration of the GABAergic system. **A** 2-deoxyglucose autoradiography shows the location of two tone representations along the extent of the CAT. The tone frequencies, used as stimuli during the experiment, differed by three octaves and activate neural representations separated by about 1 mm. **B** Schematic drawing of the section shown in **A**. Notice that the representations activated by the tone stimuli extend orthogonally to the layers of the CAT. **C** Model of the GABAergic system in the CAT in relation to the tonotopic organization. Presumed GABAergic neurons are represented by *filled symbols*. Notice that the relay cell (*open symbol*) receives input from local inhibitory neurons as well as from laterally projecting GABAergic cells. Both sets of inhibitory interneurons serve different functions of influencing the response properties of the relay cell. (from Müller and Scheich 1988). *LH*, lamina hyperstriatica; *HV*, hyperstriatum ventrale

The Organization of the GABAergic System

Anatomy

The identity of presumed GABAergic cells can be determined by antisera directed against glutamic acid decarboxylase, the GABA-synthesizing enzyme, or against GABA itself. Such selective labeling of presumed GABAergic cells allows the determination of the cell morphology, the overall number, and the distribution of such inhibitory neurons.

Using both antisera, I labeled presumed GABAergic cells in the auditory pathway of the chicken (Müller 1987a, 1988a). The data indicate that GABAergic inhibition is present throughout the entire auditory pathway, and that there are profound similarities in the distribution of GABA immunoreactive neurons and terminals in nuclei of the avian auditory pathway, and the homologue or analogue structures in the mammalian auditory system. In regard to the presence of GABAergic inhibition in field L and the adjacent hyperstriatum ventrale, the following conclusions can be drawn (Müller 1988a):

1. GABAergic inhibition is based on intratelencephalic inhibition only, because no GABAergic cells are located in the input structure (*n. ovoidalis*) of the CAT.
2. Presumed GABAergic cells are present in all layers of the CAT and contribute about 8% of the overall cell population in the auditory telencephalon.
3. Presumed GABAergic cells belong to the smallest telencephalic neurons and reveal a nonuniform morphology (bipolar to multipolar).
4. The distribution of GABAergic cells in the auditory telencephalon indicates the presence of at least two subpopulations of GABAergic cells. This conclusion is compatible with studies using Golgi impregnations, which suggest the presence of two types of presumed interneurons (with low spine density or lack of spines), i.e. local interneurons and interneurons with axons extending preferentially within one specific lamina (Saini and Leppelsack 1981).

In the mammalian visual cortex, subtypes of GABAergic interneurons have also been reported (for review, see Jones and Hendry 1986). No thalamic GABAergic cells projecting to the cortex have been described with certainty. Thus, inhibitory mechanisms seem to be only intracortical, as they are in birds. Just as in the avian telencephalon, inhibitory interneurons belong to the smallest cortical neurons and, as determined in semi-thin sections, make up a total of about 20% (Gabott and Somogyi 1986) of cortical neurons. This percentage is higher than the value obtained in the avian forebrain. However, this difference may be based on methodological differences. Recent data on thicker cortical sections reveal only 9% of GABAergic neurons in the visual cortex (Mower et al. 1988), which is virtually identical to the values obtained in my studies of thick telencephalic sections (Müller 1988b).

Like those in the telencephalon of the chicken, GABAergic cells in the mammalian cortex can be separated into those with axonal projections extending laterally and in locally acting interneurons (Gilbert 1983; Lund et al. 1988)

Physiology

In order to obtain information on the function of GABAergic inhibition in auditory telencephalic processing, I investigated the response properties of cells before and during application of the GABA-receptor antagonist bicuculline-methiodide (BIC). These studies were conducted either in fully awake birds (Müller and Scheich 1987a,b, 1988) or in an invitro slice preparation (Müller 1988b). The in vivo studies revealed the following physiological changes during microiontophoretic application of BIC:

1. The response selectively of telencephalic cells outside the input layer L2 to species-specific sounds is reduced when GABAergic inhibition is eliminated.
2. The frequency selectivity of telencephalic cells outside L2 is reduced by BIC, i.e. tuning curves are widened, or responses to pure tones are uncovered in neurons without tone responses.
3. Wide band sounds, being ineffective prior to BIC application, elicit neuronal responses after blockage of GABAergic inhibition.
4. Two-tone inhibition is partially antagonized by BIC in telencephalic cells outside layer L2.
5. Interaural inhibition is eliminated by BIC in all telencephalic layers.
6. Intensity-response curves are influenced by BIC application, i.e. inhibitory actions present at high intensity levels (response limitation) are eliminated.

Based on the topography of inhibitory actions antagonized by BIC, and with regard to activity patterns in the auditory telencephalon induced by sensory stimuli, one can distinguish two kinds of GABAergic inhibition (see Fig. 2): (a) local (within frequency) inhibition, mediating response limitation and interaural inhibition, and (b) lateral (cross-frequency) inhibition mediating the sharpening of tuning curves, two-tone suppression, and the response suppression to broad band sounds (Müller and Scheich 1988).

Those inhibitory effects that have been attributed to local interactions usually reduce spontaneous activity levels, while spontaneous activity is unchanged by presumed lateral inhibition. This difference is compatible with the assumption that local inhibition acts via synapses located close to the soma (hyperpolarizing inhibition), while lateral inhibition is mediated predominantly via dendritic synapses (shunting inhibition). Indeed, I have shown in an in vitro slice preparation that activation of local inhibition results in hyperpolarizing inhibitory postsynaptic potentials (IPSPs), while IPSPs elicited by lateral stimulation have a reversal potential positive to the cell resting membrane potential. This corresponds with the fact that dendritic application of GABA results in depolarization of cells in the chicken telencephalon, while hyperpolarizing responses are only observed with somatic application of GABA (Müller 1988b).

From the invivo experiments we could determine a lateral spread of inhibitory interactions within the auditory forebrain of a maximum of 1 mm (Müller and Scheich 1988). If we compare the data obtained in the avian auditory telencephalon with data from mammalian cortical areas (mostly obtained from cat visual cortex), striking similarities become obvious. Studies on the lateral spread of inhibition in cortical areas reveal inhibitory effects over distances as great as 1 mm (Hess et al. 1975; Asanuma and Rosen 1973). Recent data gathered using the technique of local cortical inactivation reveal disinhibitory effects of GABA application onto targets up to 1 mm from the application site (Eysel et al. 1988). The same study also shows that inactivation of sites close to a recorded neuron (within a radius of up to 500 µm) results in different disinhibitory effects than inactivation of sites further away (up to 1 mm) from a recorded neuron (Eysel et al. 1988). Both effects (i.e., reduced orientation selectively and directional selectivity of visual cortical neurons) can also be elicited by iontophoretic application of GABA-receptor antagonists onto single units (Sillito 1977, 1979) and can thus be attributed to removal of GABAergic inhibition. These data suggest that in the mammalian visual cortex, as well as in the avian CAT, local inhibition and lateral inhibition can be differentiated and serve different functions. In addition, it can be deduced that lateral inhibitory interactions are usually restricted to a maximum of 1 mm in both the avian telencephalon and the cat visual cortex.

Comparative Considerations and Open Questions

The proposed model of GABAergic inhibitory interactions in the telencephalon is based on the presence of a local and lateral inhibitory network, each of which uses a different set of interneurons and serves different functions (Müller and Scheich 1988). Different subpopulations of GABAergic cells can be characterized on the basis of their morphology (for review, see Gilbert 1983) and by presence of peptides colocalized with GABA in the mammalian cortex (Jones and Hendry 1986). Interestingly, it has been shown that colocalization of individual peptides coincides with defined sites of termination of GABAergic axons, e.g., at the soma or at dendritic segments (Jones and Hendry 1986). No such data have yet been obtained from nonmammalian species. Data on the presence (and number) of peptides colocalized with GABA will clarify the complexity of GABAergic systems in nonmammalian species.

An open question is that of what circuitry underlies lateral inhibition, which might either be mediated by laterally projecting interneurons, or lateral excitation acting on local inhibitory interneurons. In the avian auditory forebrain I have favored the former idea because presumed interneurons with laterally projecting axons are present, and because cells with frequency preferences being clearly different from the surrounding cells have not yet been found. This feature should be a characteristic of GABAergic interneurons receiving lateral excitation. However, in the mammalian cortex, combined retrograde tracing and GABA immunocytochemistry (Le Vay 1988) and combined anterograde degeneration and synaptology (Fisken et al. 1975) have not supported the presence of a dominant laterally projecting GABAergic neuronal population. Thus, it may be concluded that the similar topography of GABAergic actions is realized by different morphological substrates.

The presence of a lateral inhibitory network acting within a certain range in a functional topography allows an expanded view of the concept of central nervous modules (Szentagothai 1975). As the lateral inhibitory action can be activated from, and thus centered on virtually every point within the topography, modules can form and can be isolated from neighboring representations via a spatially limited lateral inhibition, on the basis of afferent activity rather than on intrinsic hard wiring. This concept allows the temporal formation of a limited set of modules out of a very large quantity of possible modules. This model ties together the concepts of central nervous modules (Szentagothai 1975) and neuronal assemblies (Dudai et al. 1987). Anisotropies in sensory topographies (e.g. overrepresentations of certain feature locations) might thus be viewed to allow more modules of a given size to be formed in a representation.

The Cholinergic System

Anatomy

In contrast to the wealth of information available on the organization of the cholinergic system in mammals (for review, see Fibiger 1982), only very limited information can be found on nonmammalian species. However, the available data support a high degree of similarity in cholinergic structures and projections in mammals and nonmammalian species like fish (Brantley and Bass 1988), reptiles (Brauth et al. 1985; Mufson et al. 1984), and birds (see Müller 1987b).

Physiology

I studied the action of acetylcholine (ACh) on cells in the auditory forebrain of the chicken in vivo, and in an invitro slice preparation (Müller 1987b). Two general actions of ACh can be differentiated: (1) a slow facilitatory action occurring within several seconds, and (2) a rapid inhibitory action acting within a few hundred milliseconds. These effects are topographically segregated. Inhibitory actions of ACh are observed only in the hyperstriatal parts of the CAT. The slow facilitation, however, is observed throughout the auditory nucleus, though sometimes obscured by additional inhibition. Intracellular studies, in vitro, suggest that the facilitatory action is mediated by a closure of voltage-dependent potassium channels, while the inhibition is mediated indirectly by an activation of GABAergic interneurons (Müller 1987b).

In the mammalian sensory cortex, similar observations have been made. McCormick and Prince (1986) showed in cortical slices that the inhibitory action of ACh is also mediated by an activation of GABAergic neurons. Recent iontophoretic data from the cat visual cortex, in vivo, support these findings (Müller and Singer 1989). In this study we also describe a subset of cells showing rapid excitatory responses to ACh which can be interpreted in terms that these cells are GABAergic inhibitory interneurons. As in the bird forebrain, in the mammalian cortex the inhibitory action of ACh is restricted to certain layers. Thus, cholinergic suppression is nearly exclusively observed in neurons from cortical layers II to IV, while the slow facilitation is found throughout all cortical layers (Sillito and Kemp 1983; Sato et al. 1987), again obscured by inhibition in the superficial layers (Müller and Singer 1989).

One can summarize the cholinergic actions in bird and mammalian sensory telencephalon as follows:

1. They consist of both an (indirect) inhibition and a (direct) facilitation.
2. The inhibition is restricted to certain substructures of the areas, while the facilitation is found throughout the structures.
3. The time courses of the two actions differ. The inhibition occurs very rapidly (tens of milliseconds), while the facilitation occurs after a delay of hundreds of milliseconds.

Functional Considerations

Based on the topography and the different time courses of the two cholinergic effects (suppression and facilitation), I recently proposed a model of cholinergic function (Müller 1987b, 1989; Müller and Singer 1989). It is assumed that both cholinergic suppression and facilitation are mediated by a common afferent system, and that the cholinergic system is phasically activated. According to the model (see Fig. 3) two subsequent actions of cholinergic afferents to the sensory telencephalon are distinguished:

1. A fast, short-term inhibition of those locations in the telencephalon that are part of multiple intratelencephalic recurrent loop systems. This effect leads to a rapid elimination of stabilized activity patterns in the majority of reverberating networks.
2. A delayed facilitation of telencephalic transmission, which supports the formation of new activity patterns in response to new thalamo-cortical afferent activity.

The absence of cholinergic suppression in the deep cortical layers leaves the activity in the thalamo-cortical loop system unaffected during the time of activity elimination in the intracortical loops. This might allow a selection of subsequently active cortical afferents by an interaction of the thalamic nucleus and cortical layer IV (Koch 1987; Varela and Singer 1987). Such a mechanism would be well-suited to facilitate transitions between

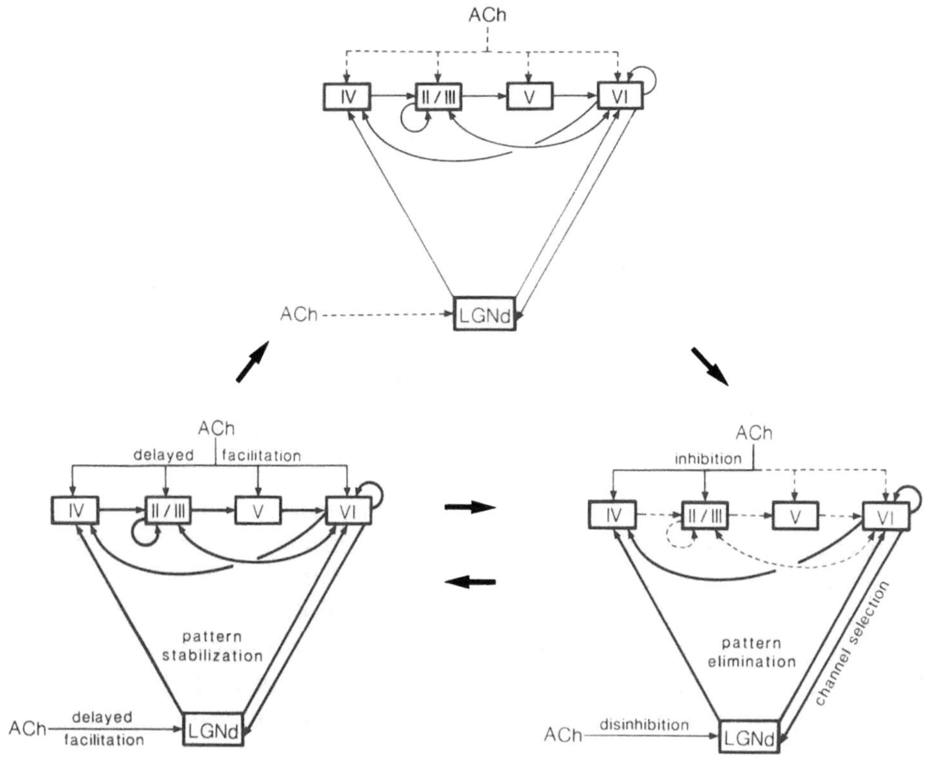

Fig. 3. Functional model of the action of cholinergic afferents to the cat visual cortex. The scheme represents the excitatory interconnectivity of the thalamo-cortical visual system; active levels carried along the projections are indicated by the lines (*broken lines*, inactive; *thin lines*, active; *thick lines*, highly active). Transient activation of the cholinergic afferents leads to a transient elimination of recurrent excitation in the cortical loop system (cholinergic suppression), followed by high activity levels (cholinergic facilitation) after decay of the suppression (see text for further information). *LGNd*, lateral geniculate nucleus, pars dorsalis; *II-VI*, cortical layers

telencephalic activity patterns in response to changing sensory input, e.g. after orienting responses. In the avian auditory system, a similar model can be constructed (Müller 1987b). However, the absence of a clear back projection from primary telencephalic sensory nuclei to the thalamus (Bonke et al. 1979a) excludes selection processes mediated by a thalamo-cortical loop system. It is conceivable that these processes are located in the telencephalic nuclei. This might explain the fact that in the avian forebrain cholinergic suppression is found only in one of the four layers, whereas neostriatal loops (between layers of field L) are unaffected by cholinergic inhibition.

Concluding Remarks

The data presented show that a comparison of the actions of two transmitter systems, in one mammalian and one nonmammalian vertebrate, reveals several functional similarities. Although there are some differences in the details of possible mechanisms (in the GABAergic system) and topographies of action (in the cholinergic system), comparable functional models can be deduced from the data. The observed structural differences might point to the presence of analogies, rather than homologies, in the ways information processing is done by the central nervous systems of different vertebrates. However, the general rules of action of the studied transmitter systems show a high degree of similarity. The proposed models set forth a framework of expectations which might be used as a guideline for future research. Finally, it is of importance to note that the similarity of integration of transmitter actions in the functional topographies of central nervous structures points to basic rules in information processing in highly evolved central nervous structures. Knowledge of these basic strategies may be of considerable importance for the understanding of the brain and in the design of artificial information processing devices.

Acknowledgements. Parts of the study have been done at the Technische Universität Darmstadt and have been supported by the DFG (SFB 45). The help of N. Gray in approving the language is kindly appreciated.

References

Asanuma H, Rosen I (1973) Spread of mono-and polysynaptic connections within cat's motor cortex. Exp Brain Res 16:507-520

Bennett TL, Nunn PJ, Inmann DP (1971) Effects of scopolamine on hippocampal theta and correlated discrimination performance. Physiol and Behav 7:451-454

Bonke BA, Bonke D, Scheich H (1979a) Connectivity of the auditory forebrain nuclei in the guinea fowl *(Numida melagris)*. Cell Tissue Res 200:101-121

Bonke D, Scheich H, Langer G (1979b) Responsiveness of units in the auditory neostriatum of the guinea fowl to species-specific calls and synthetic stimuli. I. Tonotopy and functional zones of field L. J Comp Physiol 132:243-255

Brantley RK, Bass AH (1988) Cholinergic neurons in the brain of a teleost fish *(Porichthys notatus)* located with a monoclonal antibody to choline acetyltransferase. J Comp Neurol 275:87-105

Brauth SE, Kitt CA, Price DL, Wainer BH (1985) Cholinergic neurons in the telencephalon of the reptile, *Caiman crocodilus*. Neurosci Lett 58:235-240

Cleeves L, Green SE (1982) Effects of scopolamine on visual difference thresholds in the pigeon. Physiol Psychol 10:306-312

Dewson JH, Pribram HK, Lynch JC (1969) Effects of ablation of temporal cortex upon speech sound discrimination in the monkey. Exp Neurol 24:579-591

Dudai Y, Amari SI, Bienenstock E, Dehaene S, Fuster J, Goddard GV, Konishi M, Menzel R, Mishkin M, Müller CM, Rolls ET, Schwegler HH, von der Malsburg C (1987) On neuronal assemblies and memories. In: Changeux JP, Koshini M (eds) The neural and molecular bases of learning. Wiley, New York, pp 399-410

Dykes RW, Landry P, Metherate R, Hicks TP (1984) Functional role of GABA in cat primary somatosensory cortex: shaping receptive fields of cortical neurons. J Neurophysiol 52:1066-1092

Evans HL (1975) Scopolamine effects on visual discrimination: modifications related to stimulus control. J Pharmacol Exp Ther 195:105-113

Eysel UT, Muche T, Wörgötter F (1988) Lateral interactions at direction-selective striate neurons in the cat demonstrated by local cortical inactivation. J Physiol (Lond) 399:657-675

Fibiger HC (1982) The organization and some projections of cholinergic neurons of the mammalian forebrain. Brain Res Rev 4:327-388

Fisken RA, Garey LJ, Powell TPS (1975) The intrinsic association and commissural connections of area 17 of the visual cortex. Philos Trans R Soc [Biol] 272:487-536

Gabbott PLA, Somogyi P (1986) Quantitative distribution of GABA-immunoreactive neurons in the visual cortex (area 17) of the cat. Exp Brain Res 61:323-331

Gilbert CD (1983) Microcircuitry of the visual cortex. Annu Rev Neurosci 6:217-247

Hess R, Negishi K, Creutzfeldt O (1975) The horizontal spread of intracortical inhibition in the visual cortex. Exp Brain Res 22:415-419

Hubel DH, Wiesel TN (1977) Functional architecture of macaque monkey visual cortex. Proc R Soc Lond [Biol] 189:1-59

Jenkins WM, Masterton RB (1982) Sound localization: effects of unilateral lesions in central auditory system. J Neurophysiol 47:987-1016

Jones EG, Hendry SHC (1986) Co-localization of GABA and neuropeptides in neocortical neurons. TINS 9:71-76

Koch C (1987) The action of corticofugal pathway on sensory thalamic nuclei: a hypothesis. Neuroscience 23:399-406

Le Vay S (1988) Patchy intrinsic projections in visual cortex, area 18, of the cat: morphological and immunocytochemical evidence for an excitatory function. J Comp Neurol 269:265-274

Luhmann HJ, Martinez Millan L, Singer W (1986) Development of horizontal intrinsic connections in cat striate cortex. Exp Brain Res 63:443-448

Lund JS, Hawken MJ, Parker AJ (1988) Local circuit neurons of macaque monkey striate cortex: II. Neurons of laminae 5B and 6. J Comp Neurol 276:1-29

McCormick DA, Prince DA (1986) Mechanisms of action of acetylcholine in the guinea pig cerebral cortex, in vitro. J Physiol (Lond) 375:169-194

Meadows JC (1974) The anatomical basis of prosopagnosia. J Neurol Neurosurg Psychiatry 37:489-501

Mower GD, Rustad R, White WF (1988) Quantitative comparisons of gamma-aminobutyric acid neurons and receptors in the visual cortex of normal and dark-reared cats. J Comp Neurol 272:293-302

Mufson EJ, Desan PH, Mesulam MM, Wainer BH, Levey AI (1984) Choline acetyltransferase-like immunoreactivity in the forebrain of the red-eared pond turtle (*Pseudemys* scripta elegans). Brain Res 323:103-108

Müller CM (1987a) Gamma-aminobutyric acid immunoreactivity in brainstem auditory nuclei of the chicken. Neurosci Lett 77:272-276

Müller CM (1987b) Differential effects of acetylcholine in the chicken auditory neostriatum and hyperstriatum ventrale - studies in vivo and in vitro. J Comp Physiol [A] 161:857-866

Müller CM (1988a) Distribution of GABAergic perikarya and terminals in the centers of the higher auditory pathway of the chicken. Cell Tissue Res 252:99-106

Müller CM (1988b) Function of the GABAergic inhibitory system in the chicken auditory forebrain. In: Syka J, Masterton RB (eds) Auditory system. Structure and function. Plenum, New York, pp 239-244

Müller CM (1989) Cholinergic mechanisms in the telencephalon of cat and chicken. In: Frotscher M, Misgeld U (eds) Central cholinergic synaptic transmission. Birkhäuser, Basel pp 172-182

Müller CM, Leppelsack HJ (1985) Feature extraction and tonotopic organization in the avian auditory forebrain. Exp Brain Res 59:587-599

Müller CM, Scheich H (1987a) GABAergic inhibition increases the neuronal selectivity to natural sounds in the avian auditory forebrain. Brain Res 414:376-380

Müller CM, Scheich H (1987b) GABA mediates interaural inhibition in the avian auditory forebrain. Naturwissenschaften 74:602-604

Müller CM, Scheich H (1988) Contribution of GABAergic inhibition to the response characteristics of auditory units in the avian forebrain. J Neurophysiol 59:1673-1689

Müller CM, Singer W (1989) Acetylcholine induced inhibition in the cat visual cortex is mediated by a GABAergic mechanism. Brain Res 487:335-342

Neff WD (1977) Cortical and subcortical specialization in auditory processing. In: Bullock TH (ed). Recognition of complex acoustic signals. Dahlem Konferenzen, pp 97-108 (Life sciences research reports, vol 5)

Perett DI, Rolls ET, Caan W (1982) Visual neurons responsive to faces in the monkey temporal cortex. Exp Brain Res 47:329-342

Phillis JW (1968) Acetylcholine release from the cerebral cortex: its role in cortical arousal. Brain Res 7:378-389

Saini KD, Leppelsack HJ (1981) Cell types of the auditory caudo-medial neostriatum of the starling *(Sturnus vulgaris)*. J Comp Neurol 198:209-229

Sato H, Hata Y, Masui H, Tsumoto T (1987) A functional role of cholinergic innervation to neurons in cat visual cortex. J Neurophysiol 58:765-780

Scheich H (1983) Two columnar systems in the auditory neostriatum of the chick: evidence from 2-deoxyglucose. Exp Brain Res 51:199-205

Scheich H, Langer G, Bonke D (1979) Responsiveness of units in the auditory neostriatum of the guinea fowl to species-specific calls and synthetic stimuli. II. Discrimination of jambus-like calls. J Comp Physiol 132:257-276

Sillito AM (1977) Inhibitory processes underlying the directional specificity of simple, complex and hypercomplex cells in the cat's visual cortex. J Physiol (Lond) 271:699-720

Sillito AM (1979) Inhibitory mechanisms influencing complex cell orientation selectivity and their modification at high resting discharge levels. J Physiol (Lond) 289:33-53

Sillito AM, Kemp JA (1983) Cholinergic modulation of the functional organization of the cat visual cortex. Brain Res 289:143-155

Singer W (1979) Central-core control of visual cortex functions. In: Schmitt FO, Worden FG (eds) The neurosciences fourth study program, MIT Press, Cambridge, pp 1093-1100

Sprague JM, Levy J, Di Beradino A, Berlucci G (1977) Visual cortical areas mediating form discrimination in the cat. J Comp Neurol 172:441-488

Suga N, O'Neill WE, Kujirai K, Manabe T (1983) Specificity of combination-sensitive neurons for processing of complex biosonar signals in auditory cortex of the mustached bat. J Neurophysiol 49:1573-1626

Szentagothai J (1975) The "module-concept" on cerebral cortex architecture. Brain Res 95:475-496

Varela FJ, Singer W (1987) Neuronal dynamics in the visual corticothalamic pathway revealed through binocular rivalry. Exp Brain Res 66:10-20

Winter P, Funkenstein HH (1973) The effect of species specific vocalizations on the discharge of auditory cortical cells in the awake squirrel monkey *(Saimiri sciureus)*. Exp Brain Res 18:489-504

Regional Distribution of Neurotransmitter-Related Markers: a Quantitative Microchemical Approach to the Study of Telencephalic Evolution

Antonio Contestabile[1], Rosanna Bissoli[2] and Raffaella Niso[2]
[2]Department of Animal Biology, University of Catania and
[1]Department of Biology, University of Bologna, Via Belmeloro 8, 40126 Bologna, Italy

Recent comparative studies have led to a profound revision of our concepts of the evolution of vertebrate telencephalon (for references see Northcutt 1984). In addition to embryological and neuroanatomical studies, neurochemistry is now recognized as an important tool for a better understanding of brain evolution and, in particular, for facilitating establishing homologies in different vertebrates (Reiner et al. 1984; Reiner 1986; Russchen et al. 1987). To this purpose, neurotransmitter-related markers have been considered the most valuable. Sensitive microchemical methods have been introduced which, in conjunction with microdissection techniques, allow quantitative studies of specific neurotransmitters in discrete brain regions. These methods have greatly contributed to establishing the concept of "chemical neuroanatomy" (Emson 1983).

Recently, we used this approach for a first systematic survey of major neurotransmitter systems in the main telencephalic subdivisions of representatives of each major vertebrate class. The bulk of the analytical results, as well as the methods adopted, have been published previously in a detailed form (Bissoli and Contestabile 1988; Bissoli et al. 1988, 1989). In the present paper we compare some relevant aspects of these studies.

Cholinergic Neurotransmission

Table 1 summarizes the distribution of a specific cholinergic marker, the enzyme choline acetyltransferase (ChAT), in several brain

Experimental Brain Research Series 19

Table 1. Relative ChAT values in telencephalic regions of tetrapods

	Frog	Turtle	Lizard	Pigeon	Rat
Septum					
Medial septum	97	60	33	52	87
Lateral septum	66	36	12	40	31
Striato-pallidal complex					
Striatum dorsale	84	100	84		
Striatum ventrale	82	86	100		
Paleostriatum augmentatum				100	
Paleostriatum primitivum				93	
Lobus parolfactorius				88	
Striatum					100
Globus pallidus					28
Nucleus basalis magnocellularis					39
Nucleus accumbens	100	52	47	66	80
Cortical structures					
Medial cortex	64	16	28		
Dorsal cortex	56	17	25		
Lateral cortex	60	17	21		
Hyperstriatum accessorium				32	
Hyperstriatum dorsale				29	
Prefrontal cortex					27
Motor cortex					20
Primary sensory cortex					19
Primary olfactory cortex					47
Primary visual cortex					20
Accessory visual cortex					20
Cingulate cortex					26
Hippocampus					26
DVR and derivates					
Anterior DVR		52			
Lateral anterior DVR			19		
Central anterior DVR			24		
Medial anterior DVR			18		
Hyperstriatum ventrale anterior				33	
Neostriatum laterale anterior				12	
Ectostriatum				4	
Neostriatum				28	
Field L area				6	

DVR, dorsal ventricular ridge. Species of animals used: Frog, *Rana esculenta*; Turtle, *Pseudemys scripta elegans*, Lizard, *Anolis caroliniensis*, Pigeon *Columba livia* (King); Rat, *Rattus norvegicus* (Wistar).

regions of tetrapods. In this, as well in the following tables, brain regions have been listed in homogenous in homogenous groups so as to facilitate their comparison. We tried, however, to maintain a traditional and neutral terminology (we therefore refer, for instance, to dorsal and ventral striatum for reptiles and paleostriatum augmentatum and primitivum in birds instead of using the terms striatum and globus pallidus). The results have been expressed giving the value of 100 to the area with the highest level of the marker for each species; the values of the other areas have been expressed as a percentage of this value. This makes it easier to immediately compare the levels of the markers in terms of the ratio between different areas rather than in terms of the absolute values, which can be very different from animal to animal. The following main points emerge from the regional distribution of the cholinergic marker in the vertebrates studied.

1. A common basic pattern is represented by the higher cholinergic levels of the basal telencephalon (in particular of the striatum-accumbens complex) in comparison with cortical structures. This feature is barely evident in the frog, but it is obvious in amniotes.
2. Among basal telencephalic areas some degree of compartmentalization exists between the higher cholinergic levels of the medial septum-diagonal band complex and the lower levels of the lateral septum. A strong compartmentalization between striatal and pallidal components is evident in mammals but not in other tetrapods.
3. The level of ChAT in the anterior dorsal ventricular ridge (DVR) of the turtle lies between the values of the striatum and those of the cortices, while in the iguanid lizard ChAT levels in the three subdivisions of the anterior DVR and of the cortices are essentially the same.
4. The cholinergic levels in the hyperstriatum accessorium and dorsale, as well as in some of the DVR derivates of the pigeon, are comparable with those found in the rat neocortex on the basis of the ratio with the respective striatal values. By contrast, the ectostriatum and the area including the neostriatal "field L" are characterized by extremely low ChAT levels.

In several cases our neurochemical results do not correspond to the reported distribution of muscarinic nicotinic receptors obtained by autoradiography in birds (Dietl et al. 1988a; Watson et al. 1988). The reasons for these apparent discrepancies cannot be discussed in detail here. However, similar mismatch situations have been often described by mammalian neurochemists.

Immunocytochemical studies in mammals (Mesulam et al. 1983) have demonstrated the existence of two main populations of cholinergic neurons in the basal telencephalon: interneurons in the striatum and projection neurons distributed from the medial septum to the nucleus basalis. The latter neurons are responsible for the

Table 2. Relative ChAT values in the telencephalon of dogfish *(Scyliorhinus canicula)* and goldfish *(Carassius auratus)*.

	Dogfish	Goldfish
Presumptive basal areas		
Septum	75	
Striatum	54	
Superficial basal area	59	
Ventral area		100
Dorsomedial area		93
Dorsocentral area		44
Presumptive pallial areas		
Dorsal pallium, pars centralis	83	
Dorsal pallium, pars superficialis	55	
Medial pallium	100	
Lateral pallium	60	
Dorsodorsal area		50
Dorsolateral area		37

Subdivision of basal and pallial telencephalic areas has been, in part, done following the ideas expressed by Northcutt and Braford (1980) for teleosts. The data for the dogfish are the results from only four experiments; for this reason the standard errors of the means are rather high and the apparent differences are not very significant statistically. It is, however, clear that the pattern of ChAT distribution noticed for tetrapods is not present in the dogfish since pallial areas possess cholinergic levels equal or even higher than basal areas.

bulk of the cholinergic innervation to the cortex and hippocampus. As a consequence of the extensive local innervation from cholinergic interneurons, ChAT levels in the striatum and nucleus accumbens largely exceed those measured in the globus pallidus (Fonnum and Walaas 1979). As far as nonmammalian tetrapods are concerned, data are only available for reptiles. In *Caiman* (Brauth et al. 1985), cholinergic cell bodies are localized in areas apparently corresponding to those of mammals with a lower density in the striatum proper. In the turtle, however, the striatal component of cholinergic neurons seems completely lacking (Mufson et al. 1984). No data on ChAT localization have been published so far for birds; however, results of acetylcholinesterase histochemistry suggest a distribution of cholinergic neurons more similar to that of the turtle (Parent 1986). A possible interpretation of these data, as well as of the absence of clear-cut differences in cholinergic levels between striatal and pallidal areas, is that in nonmammalian tetrapods the cholinergic neurons of the basal telencephalon profusely innervate the entire striato-pallidal complex. It is also likely that a subpopulation of these neurons gives rise to cholinergic innervation and dorsal telencephalic areas (Mufson et al. 1984). In mammals, an anatomical and functional subdivision of these cholinergic neurons could have occurred between the striatal interneurons and the projection neurons of the globus pallidus and nucleus basalis. Indirect evidence for such a process comes from physiological investigations which demonstrate some basic differences in the interactions between cholinergic and other neurotransmitter systems in reptiles and mammals (Stoof et al. 1987). The compartmentalization we have observed in the cholinergic levels between the medial septum-diagonal band region and the lateral septum seems to correlate well with the preferential localization of cholinergic neurons in the medial subdivision of all vertebrates so far investigated.

Table 2 lists the levels of ChAT in telencephalic regions of a cartilaginous fish and a teleost. No relevant regional differences are apparent in the dogfish, while in the goldfish the ventral and the dorsomedial area display levels 2-2.5 times higher than other regions. Immunocytochemistry in teleosts has revealed the presence of a small cholinergic population of cells localized in the ventral area (Brantley and Bass 1988). While the possibility that cholinergic projections may come from extratelencephalic sources cannot be excluded, this population may be wholly responsible for telencephalic cholinergic innervation. If so, a situation similar to that hypothesized for nonmammalian tetrapods holds true for teleosts.

GABAergic Neurotransmission

Table 3 summarizes the quantitative distribution of a specific GABAergic marker, the enzyme glutamate decarboxylase (GAD), in brain regions of tetrapods. GABA is one of the most diffuse neurotransmitters and GABAergic neurons have been estimated to represent a substantial proportion (more than 30%) of the total neuronal population in several brain areas of mammals (McGeer et al. 1978). Since most GABAergic neurons are intrinsic interneurons, the levels of GAD are usually considered to be correlated with the density of these neurons in a given region. This may explain the absence, in most cases, of dramatic differences in the regional levels of GABAergic markers. There are, however, examples of GABAergic connections between different nuclei, and in this case GABAergic markers in the recipient area may reach levels several times higher

Table 3. Relative GAD values in telencephalic areas of tetrapods

	Frog	Turtle	Lizard	Pigeon	Rat
Septum					
Medial septum	72	48	50	67	100
Lateral septum	64	26	21	50	76
Striato-pallidal complex					
Striatum dorsale	70	39	33		
Striatum ventrale	100	100	100		
Paleostriatum augmentatum				41	
Paleostriatum primitivum				100	
Lobus parolfactorius				62	
Striatum					31
Globus pallidus					98
Nucleus basalis magnocellularis					100
Nucleus accumbens	89	53	49	82	56
Cortical structures					
Medial cortex	39	21	26		
Dorsal cortex	23	20	29		
Lateral cortex	32	21	31		
Hyperstriatum accessorium				25	
Hyperstriatum dorsale				25	
Prefrontal cortex					36
Motor cortex					39
Primary sensory cortex					45
Primary olfactory cortex					43
Primary visual cortex					42
Accessory visual cortex					43
Cingulate cortex					43
Hippocampus					44
DVR and derivates					
Anterior DVR		27			
Lateral anterior DVR			21		
Central anterior DVR			25		
Medial anterior DVR			29		
Hyperstriatum ventrale anterior				23	
Neostriatum laterale anterior				27	
Ectostriatum caudale				23	
Neostriatum caudale				27	
Field L area				32	

GAD, glutamate decarboxylase; *DVR*, dorsal ventricular ridge

than the average brain value. This is, for instance, the case for the GABAergic striato-pallidal connection in mammals which raises the pallidal level of GAD up to a level more than three times higher than in the striatum (Fonnum and Walaas 1979). One finds similar differences in the GAD levels if one compares the ventral (pallidal) and dorsal (striatal) components of the striatal complex of reptiles and birds. The nucleus accumbens in all cases shows intermediate values (Table 3). A similar, but less striking regional compartmentalization seems to be present in amphibians. A second, rather constant feature in tetrapods is a small, but definitive difference in the levels of GAD between the medial septum-diagonal band complex (higher) and the lateral septum (lower). Recently published results on the distribution of GABA receptors in the pigeon brain (Dietl et al. 1988b) agree in general with our data.

In cartilaginous and teleost fish (Bissoli et al. 1989) no great differences in the levels of GAD exist among the various telencephalic areas listed in Table 2.

Excitatory Amino Acid Neurotransmission

The main way to terminate the effect of the released excitatory amino acids glutamate and aspartate is by re-uptake into the terminal from the synaptic cleft (Fonnum 1984). The D isomer of aspartate is recognized and transported by the same carrier without being further metabolized, and its high affinity uptake from synaptosomes in vitro is, therefore, a reliable marker for the density of terminals releasing excitatory amino acids in a given brain area. The neurotransmission mediated by excitatory amino acids is reputed to be quantitatively very important in mammals (Fonnum 1984). Our own studies (Bissoli and Contestabile 1988; Bissoli et al. 1988, 1989) have revealed that excitatory amino acids play a major role in the telencephalon of all vertebrates since in most cases we have measured levels of uptake comparable to those of mammals.

Table 4 summarizes the regional levels measured in tetrapods. These data may suggest that levels of uptake are higher in cortical than in basal telencephalic areas. Furthermore, in mammals a very active uptake is measured in the striatum, reflecting the presence of a massive cortico-striatal glutamatergic input (Fonnum 1984). In contrast, the globus pallidus displays an extremely low uptake. A similar, although less striking compartmentalization in striatal and pallidal areas is present in reptiles and birds, whereas it is absent in amphibians. In telencephalic areas of cartilaginous and teleost fish, a rather uniform distribution has been observed (Bissoli et al. 1989).

Table 4. Relative D - 3H aspartate values in telencephalic areas of tetrapods

	Frog	Turtle	Lizard	Pigeon	Rat
Septum					
Medial septum	47	39	64	80	15
Lateral septum	50	43	48	90	30
Striato-pallidal complex					
Striatum dorsale	38	46	98		
Striatum ventrale	40	25	60		
Paleostriatum augmentatum				57	
Paleostriatum primitivum				32	
Lobus parolfactorius				78	
Striatum					79
Globus pallidus					6
Nucleus basalis magnocellularis					14
Nucleus accumbens	51	38	78	90	63
Cortical structures					
Medial cortex	80	62	63		
Dorsal cortex	81	100	63		
Lateral cortex	100	89	63		
Hyperstriatum accessorium				83	
Hyperstriatum dorsale				100	
Prefrontal cortex					100
Motor cortex					84
Primary sensory cortex					50
Primary olfactory cortex					28
Primary visual cortex					54
Accessory visual cortex					61
Cingulate cortex					52
Hippocampus					61
DVR and derivates					
Anterior DVR		90			
Lateral anterior DVR			74		
Central anterior DVR			56		
Medial anterior DVR			54		
Hyperstriatum ventrale anterior				99	
Neostriatum laterale anterior				85	
Ectostriatum caudale				65	
Neostriatum caudale				92	
Field L area				83	

DVR, dorsal ventricular ridge

Comparative Considerations

Theoretical and Technical Problems

An important question is: to what extent is one allowed to consider the results obtained in just one or two species as representative of a whole vertebrate class? To answer this question let us first consider the situation of mammals in which neurochemical data are available for a rather wide range of species. These data show that the relative levels in corresponding regions are remarkably constant in the various species. Data are fewer in number and somewhat contrasting for nonmammalian vertebrates. Different regional distributions for the same neurotransmitter in different species belonging to the same class have been reported in teleost telencephalon for ChAT (Bissoli et al. 1989) and in the DVR of reptiles for ChAT (present data) and dopamine (Russchen et al. 1987; Smeets et al. 1987). However, other data on the distribution of ChAT, dopamine, and several neuropeptides suggest that no major differences exist in the basic pattern of regional neurotransmitter distribution between different species belonging to the same class (Inagaki et al. 1981; Taban and Cathieri 1983; Reiner et al. 1984; Parent 1986; Russchen et al. 1987).

In view of the whole body of available information, it is our opinion that the study of regional neurochemistry in telencephalic areas of selected vertebrates may tell us something about the general pattern of evolution of the neurotransmitter systems, even though particular structures may require individual attention (this seems to be the case for the DVR) and, in some cases, technical problems (for instance sensitivity limits of histochemical methods or difficulties in reproducible microdissection of very small areas) may complicate the interpretation of the results.

Evolution of Neurotransmitter Systems

In cartilaginous fish the three neurotransmitter systems studied are uniformly distributed throughout the telencephalic areas. In the frog, a process of neurochemical differentiation between basal and cortical telencephalic areas begins; this differentiation becomes progressively more evident in amniotes. Regional compartmentalization for GABA between striatal and pallidal areas is poor in amphibians but evident in amniotes. In the same region, cholinergic compartmentalization is obvious in mammals only. Compartmentalization for excitatory amino acids is absent in amphibians, poor in reptiles and birds, but very strong in mammals.

We have observed comparable features for cholinergic and GABAergic compartmentalization between the medial septum-diagonal band complex and the lateral septum. In addition, several histochemical data suggest comparable trends of neurochemical compartmentalization in tetrapods, e.g., the regional distribution of substance P, enkephalins, and catecholamines in the striato-pallidal complex (Inagaki et al. 1981; Tabañ and Cathieri 1983; Reiner et al. 1984; Parent 1986; Franzoni et al. 1986; Russchen et al. 1987). Taken together, these data suggest that evolution of neurotransmitter systems has involved a process of progressive regional compartmentalization starting from a primitive condition characterized by a relatively undifferentiated and diffuse distribution.

How does this presumed pattern fit into theories of brain evolution? Since various neurotransmitter systems are present in diffuse form in the primitive condition, it is not easy to imagine that neurochemical regional specialization should be mainly due to colonization of new fiber systems as postulated by the so-called invasion hypothesis. The evolutionary trend observed appears to be better explained by a combination of the parcellation theory (Ebbesson 1980) and the equivalent cell hypothesis (Nauta and Karten 1970). The parcellation theory postulates that regional differentiation is the result of selective loss of some connections and the concomitant strengthening of some of the remaining ones in a given brain area. The equivalent cell hypothesis considers cell populations homologous that are characterized by similar connections and neurochemical properties. A corollary of this theory is that homologous cell populations are not necessarily found in the same topographical localization in different species since they can migrate from their region of origin, still keeping their original connections and neurochemical characters. The process of neurochemical compartmentalization first between dorsal and basal telencephalon and then between striatal and pallidal areas successively, seems a good paradigm to test the validity of these theories since both loss of connections (monitored, for instance, by the decrease of cholinergic levels in the dorsal telencephalon as compared with the basal one and in the globus pallidus as compared to the striatum) and cell migration (monitored by the rearrangement of cholinergic and peptidergic neurons in the striato-pallidal complex) should have occurred. A case in which the invasion hypothesis could be invoked is the progressive increase of excitatory amino acidic transmission in the striatum. This may be ascribed to the appearance and increase of glutamatergic cortico-striatal input in amniotes since it was believed that no cortico-striatal connections were present in amphibians (Wilczynski and Northcutt 1983). However, the recent finding of an incipient cortico-striatal connection in amphibians (see Fasolo et al., this volume) suggests that what happens may not be the invasion of the striatum by a new pathway, but the progressive strengthening of a preexisting one.

In teleosts, characterized by an everted type of telencephalic development, regional neurochemical differentiation seems in many cases absent. The only relevant exception we have noticed is the high ChAT level in the ventral and dorsomedial area of the goldfish. On the basis of anatomical and histochemical data (Northcutt and Braford 1980), it has been suggested that these areas may be homologous to septum-amygdala and striatum, respectively. Thus, also in teleosts regional neurochemical differentiation may be expressed, at some evolutionary stages, by features essentially similar to those reached in the evolutionary line characterized by the inverted telencephalon development.

The Problem of the Dorsal Ventricular Ridge

The definition of the homology of the DVR with other telencephalic areas is a central point for better understanding telencephalic evolution in tetrapods. On the basis of embryological, histochemical and anatomical studies, a rather large consensus exists in that the DVR may be considered a neocortical equivalent (Källén 1951, Ulinski 1983, Parent 1986). Neurochemical data, however, suggest that in some cases the DVR exhibits features more similar to the striatum than to the cortex. This is, for instance, the case for the cholinergic levels in the turtle DVR (present data), the high dopamine content of the lizard DVR (Russchen et al. 1987) and the density of dopamine receptors in the turtle DVR (Richfield et al. 1987). We suggest that these "striatal" neurochemical features mark a peculiar transitional stage of DVR evolution, characterized by the presence, together with neuronal populations of pallial origin (Källén 1951), of neuronal populations of striatal origin which migrate dorsally into the DVR. Embryological evidence in support of the above-proposed process is, at present, lacking. That a migration of striatal neurons towards the DVR may have occurred is suggested by the fact that in amphibians the striatum receives its heaviest input from sensory thalamic nuclei relaying midbrain auditory and visual modalities (Wilczynski and Northcutt 1983) and that the same inputs are relayed to specific areas of the DVR in reptiles and birds. If future research is able to precisely correlate the "striatal" neurochemical features of the DVR with "striatal" anatomical connections (demonstrating, for instance, a dopaminergic projection to the DVR from the reptilian homologue of the substantia nigra), the above-proposed hypothesis will be significantly supported. Further evolution of the DVR is unique to this structure and most likely involves large rearrangement with increase of some neuronal populations, relative decrease of others, and drastic changes in some aspects of synaptic neurochemistry. Such processes are likely at the basis of some of our observations, like those concerning the dramatic decrease of the cholinergic levels in areas of the bird DVR supposedly homologous to mammalian accessory visual cortex and auditory cortex.

The hypothesis on the evolution of the DVR we have expressed here is not in agreement with current ideas about the problem. However, we believe that some facts contrasting with a total homologization of the DVR with neocortex, such as some discrepancies in synaptic neurochemistry and the fact that DVR does not apparently project outside the telencephalon, are now sufficient to propose partially alternative hypotheses and to devise experimental approaches to test them.

Acknowledgements. The studies dealt with in this article have been supported by grants from the Italian Ministry of Education.

References

Bissoli R, Contestabile A (1988) Evolution of neurotransmitter markers in vertebrate telencephalon. Comparative microchemical study in discrete brain regions of a frog and a turtle. Comp Biochem Physiol 89C:241-248

Bissoli R, Battistini S, Guarnieri T, Contestabile A (1988) Regional levels of neurotransmitter markers in the pigeon telencephalon: a comparison with possibly homologous areas of the rat telencephalon. J Neurochem 59:1731-1737

Bissoli R, Contestabile A, Niso R, Szabo T (1989) Regional levels of cholinergic, GABAergic and excitatory amino acidic transmitters in fish telencephalon. Comp Biochem Physiol (in press)

Brantley, RK, Bass AH (1988) Cholinergic neurons in the brain of a teleost fish *(Porichthys notatus)* localized with a monoclonal antibody to choline acetyltransferase. J Comp Neurol 275:87-105

Brauth SE, Kitt CA, Price DL, Wainer BH (1985) Cholinergic neurons in the telencephalon of the reptile, *Caiman crocodilus*. Neurosci Lett 58:235-240

Dietl MM, Cortés R, Palacios JM (1988a) Neurotransmitter receptors in the avian brain. II. Muscarinic cholinergic receptors. Brain Res 439:360-365

Dietl MM, Cortés R, Palacios JM (1988b) Neurotransmitter receptors in the avian brain. III. GABA-benzodiazepine receptors. Brain Res 439:366-371

Ebbesson SOE (1980) The parcellation theory and its relation to interspecific variability in brain organization, evolutionary and ontogenetic development and neuronal plasticity. Cell Tissue Res 213:179-212

Emson PC (1983) Chemical Neuroanatomy. Raven, New York

Fonnum F (1984) Glutamate: a neurotransmitter in mammalian brain. J Neurochem 42:1-11

Fonnum F, Walaas I (1979) Localization of neurotransmitter candidates in neostriatum. In: Divac I, Öberg RGE (eds) The neostriatum. Pergamon, New York, pp 53-69

Franzoni MF, Thibault J, Fasolo A, Martinoli MG, Scaranari F, Calas A (1986) Organization of tyrosine hydroxylase immunopositive neurons in the brain of the crested newt *Triturus cristatus carnifex*. J Comp Neurol 251:121-134

Inagaki S, Senba E, Shiosaka S, Takagi H, Kawai Y, Takatsuki K, Sakanaka M, Matsuzaki T, Toyama M (1981) Regional distribution of substance-P-like immunoreactivity in the frog brain and spinal cord: immunohistochemical analysis. J Comp Neurol 201:243-254

Källén B (1951) On the ontogeny of the reptilian forebrain. Nuclear structures and ventricular sulci. J Comp Neurol 95:307-347

McGeer PL, Eccles JC, McGeer EG (1978) Molecular neurobiology of the mammalian brain. Plenum, New York

Mesulam MM, Mufson EJ, Wainer BH, Levey AI (1983) Central cholinergic pathways in the rat: an overview based on an alternative nomenclature (Chl-Ch6). Neuroscience 10:1185-1201

Mufson EJ, Desan PH, Mesulam MM, Wainer BH, Levey AI (1984) Choline acetyltransferase-like immunoreactivity in the forebrain of the red-eared pond turtle *(Pseudemys scripta elegans)*. Brain Res 323:103-108

Nauta WJH, Karten HJ (1970) A general profile of the vertebrate brain, with sidelights on the ancestry of cerebral cortex. In: Quarton GC, Melnechuk T, Schmitt FO (eds) The neuroscience second study program. Rockefeller University Press, New York pp 7-26

Northcutt RG (1984) Evolution of the vertebrate central nervous system: patterns and processes. Am Zool 24:701-716

Northcutt RG, Braford MR (1980) New observations on the organization and evolution of the telencephalon of actinopterygian fishes. In: Ebbesson SOE (ed) Comparative neurology of the telencephalon. Plenum, New York, pp 41-98

Parent A (1986) Comparative neurobiology of the basal ganglia. Wiley, New York

Reiner A (1986) Is prefrontal cortex found only in mammals? Trends Neurosci 9:298-300

Reiner A, Brauth SE, Karten HJ (1984) Evolution of the amniote basal ganglia. Trends Neurosci 7:320-325

Richfield EK, Young AB, Penney JB (1987) Comparative distribution of dopamine D-1 and D-2 receptors in the basal ganglia of turtles, pigeons, rats, cats and monkeys. J Comp Neurol 262:446-463

Russchen FT, Smeets WJAJ, Hoogland PV (1987) Histochemical identification of pallidal and striatal structures in the lizard *Gekko gekko*. Evidence for compartmentation. J Comp Neurol 256:329-341

Smeets WJAJ, Jonker AJ, Hoogland PV (1987) Distribution of dopamine in the forebrain and midbrain of the red-eared turtle, *Pseudemys scripta elegans*, reinvestigated using antibodies against dopamine. Brain Behav Evol 30:121-142

Stoof JC, Russchen FT, Verheijden PFHM, Hoogland PV (1987) A comparative study of dopamine-acetylcholine interaction in telencephalic structures of the rat and of a reptile, the lizard *Gekko gekko*. Brain Res 404:273-291

Taban CH, Cathieri M (1983) Distribution of substance P-like immunoreactivity in the brain of the newt *(Triturus cristatus)*. J Comp Neurol 216:453-470

Ulinski PS The Dorsal Ventricular Ridge. A treatise on forebrain organization in reptiles and birds. Wiley, New York

Watson JT, Adkins-Regan E, Whiting P, Lindstrom JM, Podleski TR (1988) Autoradiographic localization of nicotinic acetylcholine receptors in the brain of the zebra finch *(Poephilia guttata)*. J Comp Neurol 274:255-264

Wilczynski W, Northcutt RG (1983) Connections of the bullfrog striatum: afferent organization. J Comp Neurol 214:321-332

Archicortical and Periarchicortical Areas in the Vertebrate Forebrain

Walter K. Schwerdtfeger and Peter Germroth
Max-Planck-Institut für Hirnforschung,
Deutschordenstr. 46, D-6000 Frankfurt/M., FRG

The archicortical and periarchicortical parts of the mammalian brain comprise the hippocampal (with Ammon's horn, area dentata and subiculum), entorhinal, presubicular and retrosplenial regions (Filimonoff 1947, Stephan 1975). Whether such areas also exist in the brain of nonmammalian vertebrate species has been discussed for about one century (for review of earlier literature, see Stephan 1975). The controversy persists since some neurobiologists are more than skeptical about the attempts to find homologies between the pallial structures of mammals and those of other vertebrates. The existence of such homologies, however, can only be doubted if one assumes a polyphyletic development of the vertebrate classes. If one accepts, like most authors, that the terrestrial vertebrates have evolved from one common ancestor group of fishes, one should also assume that their brains are derived from a common forerunner brain. If the latter contained a telencephalon, it also possessed the structure(s) that developed into archicortex and periarchicortex during the course of evolution. There is paleontological evidence of the early appearance of the telencephalon, which already existed in the brains of fossil Acrania and fishes (Starck 1978).

Comparative neuroanatomical studies indicated a close correlation between the development of a given brain structure and the functional importance of the latter in the respective species (e.g. Stephan et al. 1988 and references therein). The basic organization (*Bauplan*) of all vertebrate species is similar, as are the principal tasks to be performed by a central nervous system, such as the control of alimentation, orientation, reproduction and internal regulation. However, the degree of differentiation and size of the structure may differ greatly even between closely related species according to their different biological habits and adaptations (Stephan et al. 1988).

Since some structures in certain species may have maintained original features of organization, the comparative study of subdivisions of the vertebrate brain aims not only at the

Experimental Brain Research Series 19

phylogenetic relationships but also at general principles of central nervous system organization.

Terminology

The embryonic telencephalic walls of vertebrates comprise medial, dorsal, lateral and basal parts. The former three constitute the pallium, the latter the basal nucleus (e.g. Smith 1901; Kuhlenbeck 1929). In the mammalian brain, the pallium gives rise to the archicortex, neocortex and part of the paleocortex (the prepiriform cortex or true paleocortex since it develops from a cortical plate; Angevine and Sidman 1962), whereas the basal nucleus differentiates into the remainder of paleocortex and the basal ganglia.

Like Kuhlenbeck (1929), we define the cortex as the gray substance situated immediately below the telencephalic surface and separated from the periventricular matrix/ependyma by a distinct fiber layer that does not contain neurons. According to this definition, true cortices are only formed in the brain of reptiles and mammals. A rough but useful subdivision of the cortical structures is based on their afferent input, especially their relation to the sensory systems: paleocortex receives secondary olfactory fibers, i.e. direct projections from the olfactory bulb but no secondary and tertiary fibers from other sensory systems; neocortex receives from the thalamus tertiary sensory fibers of nonolfactoriy origin; archicortex receives tertiary olfactory fibers (possibly also secondary olfactory fibers in the anterior hippocampal rudiment, and tertiary visual fibers). Paleocortical and archicortical areas have no or only faint connections with neocortex. Certainly some transitional zones exist (peripaleocortex, periarchicortex), such as the entorhinal cortex. Here, secondary and tertiary olfactory fibers terminate but the most important connections are those with the neocortex and archicortex.

Cytoarchitecture

On the basis of cytoarchitectural comparisons some areas in the brains of fishes, amphibians, reptiles and birds have been delimited which may be homologous with mammalian archicortex and periarchicortex (Fig. 1). They are descendants of the medial wall

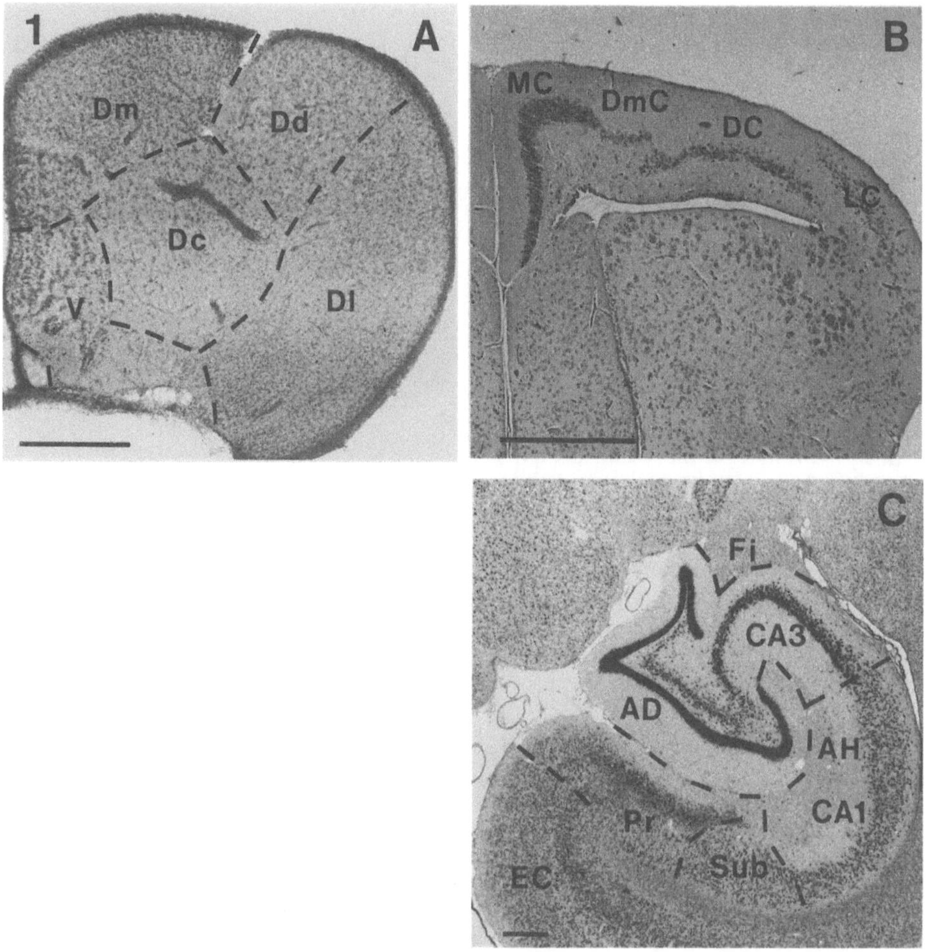

Fig. 1 A-C. Subdivisions of the telencephalon in goldfish (*Carassius auratus*, **A**) and lizard (*Podarcis hispanica*, **B**), and of archicortical and adjacent periarchicortical areas in a monkey (*Callithrix jacchus*, **C**). *AD*, area dentata; *AH*, Ammon's horn with fields CA1 and CA3; *DC*, dorsal cortex; *Dc, Dd, Dl, Dm*, central, dorsal, lateral and medial parts of dorsal pallium; *DmC*, dorsomedial cortex; *EC*, entorhinal cortex; *Fi*, fimbria; *LC*, lateral cortex; *MC*, medial cortex; *Pr*, presubicular region; *Sub*, subiculum; *V*, ventral pallium. Nissl staining; *Scale bars*, 0.5 mm

and of parts of the dorsal and lateral walls of the embryonic telencephalon. In elasmobranchs, dipnoans and amphibians, three principal longitudinal zones are recognized, namely, the lateral, dorsal and medial pallium (e.g. Kuhlenbeck 1927), of which the latter two have been termed general pallium (Johnston 1924) and *primordium hippocampi* (Smith 1903), respectively. The same zones exist in the brains of ganoids, teleosts and *Latimeria*, although

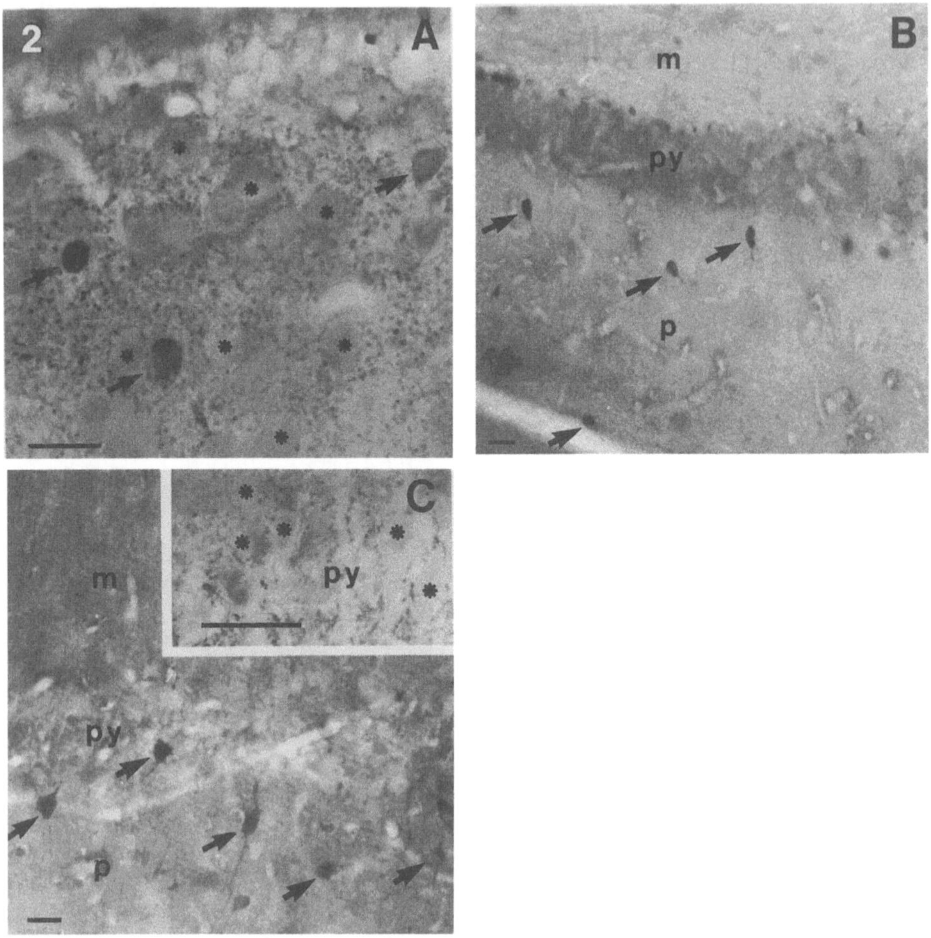

Fig. 2 A-C. GABA-immunoreactive neurons (*arrows*) in the dorsal pallium (dorsal part) of the goldfish (**A**), dorsal cortex of the lizard (**B**), and hippocampus (field CA1) of the rat (**C**), with *inset* showing part of the pyramidal cell layer. Somata of unstained neurons (some marked by *) are encircled by GABA-positive axonal boutons. *m*, molecular layer; *p*, polymorphous layer; *py*, pyramidal layer. *Scale bars*, 25 μm

possibly in an everted arrangement (for discussion see Northcutt and Braford 1980) whereby the primordium hippocampi occupies a lateral, instead of a medial, position in the forebrain. However, this point is still a matter of controversial discussion (cf. Northcutt and Braford 1980; Schnitzlein 1982; Murakami et al. 1983).

The amphibian medial pallium displays medial and dorsomedial longitudinal zones (Northcutt and Kicliter 1980). In reptiles, especially the Squamata, the corresponding region is subdivided into medial and dorsomedial cortices (Ebbesson and Voneida 1969; Ulinski 1977) that appear rather similar to the mammalian hippocampus since they are distinctly laminated into molecular, cellular and polymorphous layers with granule cell-like (medial cortex) and pyramid-like (dorsomedial cortex) neurons dominating the cellular layer. Nonsquamate reptiles show a less clear medial/dorsomedial parcellation (cf. Smeets et al. 1986a, lizard; Desan 1988, turtle). In birds, finally, a hippocampus is generally delimited (e.g. Rehkämper et al. 1988), although its lamination and areal borders are less distinct than in mammals. Likewise, a caudal part of the *hyperstriatum accessorium* has been termed parahippocampus by some authors (see Kuhlenbeck 1977).

Cytology

One finds two basic types of pallial neurons throughout the vertebrate scale, namely the spinous neurons and the sparsely spinous and aspinous neurons. In terms of function, nearly all of the spinous neurons are excitatory, whereas aspinous and sparsely spinous cells are inhibitory, most of them using the neurotransmitter GABA. In the nonlaminated pallium of fishes and amphibians, the two types are more or less homogenously distributed throughout; in the distinctly laminated cortices of reptiles and mammals most of the spinous cells are concentrated in one or more distinct principal cell layer(s), and the majority of the sparsely spinous and aspinous cells are found in the outer (molecular) and inner (polymorphous) layers (Figs. 1, 2). In the nonlaminated pallium and in the subpallial/subcortical forebrain nuclei of all vertebrate classes, dendrites of most spinous neurons are arranged in a multipolar matter and their somata are spherical to ovoid. Concomitant with the development of cortical lamination paralleled by lamination of afferent input, the spinous cells assume a vertical orientation with their generally vertically oval or pyramid-like somata (pyramidal cells) located in the principal cell layer(s) and their apical dendrites ascending towards the pia. Besides the pyramidal cells, stellate and horizontal spinous neurons have been described. In mammals, stellate cells occur in the outer cell layer (layer II) of periarchicortex (Lorente de Nó 1933), paleocortex (Valverde 1983) and neocortex (Lund 1984). Similar neurons have been found in pallial areas of amphibians (Northcutt and Kicliter 1980), reptiles (Ebbesson and Voneida 1969) and birds (see Tömböl, this volume). An example of horizontal spinous neurons are the mossy cells of the mammalian hippocampus (Amaral 1978). Nearly all spinous cortical neurons extend a long, projecting axon that forms asymmetric synaptic contacts on the somata and dendrites of aspinous and sparsely spinous neurons and on the dendrites of other spinous cells, whereas their soma receives nearly exclusively symmetric contacts (Peters and Jones 1984; Lund 1984).

The aspinous and sparsely spinous neurons of the pallium constitute a heterogenous cell population of generally nonpyramidal appearance since their somata are unlike pyramids and their main dendrites are arranged in multipolar, horizontal-bipolar and vertical-bipolar manners. A notable exception is constituted by the pyramidal basket cells, neurons with predominantly vertical-oval somata and one ascending apical dendrite, which occur in the cortices of mammals (Lorente de Nó 1934) and, probably, of reptiles (Schwerdtfeger and Lorente 1988). The axons of the aspinous and sparsely spinous neurons are generally short and intensely branched, making symmetric synaptic contacts mainly with the somata and dendrites of spinous neurons (see Fig. 2). However, synapses are also formed on other nonpyramidal cells (Misgeld and Frotscher 1986; Schwerdtfeger and Lorente 1988). One type of sparsely spinous neurons, the chandelier cells, contact exclusively the axon initial segment of spinous neurons (Somogyi 1979).

Generally, all of these synaptic contacts are formed within the area in which the soma of the aspinous/sparsely spinous cell is located. Thus these cells have been designated interneurons or local-circuit neurons, in contrast to the spinous neurons which have been termed projection neurons. There are several reasons why the former classification of pallial neurons needs some revision: (1) The application of refined techniques, such as a combination of retrograde tracing and Golgi staining or intracellular filling of retrogradely labeled cells, has made clear that many nonpyramidal neurons extend a long, projecting axon (see Schwerdtfeger and Buhl 1986 and references therein; Buhl et al. 1989; Germroth et al. 1989); (2) In mammalian neocortex, pyramidal neurons have been found that are characterized by an intrinsic axon (Lund and Boothe 1975), and there is electrophysiological evidence that pyramidal neurons extend axon collaterals acting in local excitation of other excitatory neurons (McVicar and Dudek 1980); (3) At least one type of cortical spinous neurons is GABAergic (Misgeld and Frotscher 1986; Schwerdtfeger and Lorente 1988).

Thus, at present, one can discern excitatory and inhibitory neurons in the vertebrate pallium, bearing in mind that both of them may contact local, as well as distant, targets.

It is possible that the number of aspinous and sparsely spinous inhibitory neurons in the pallium has increased during phylogeny. The paleopallium contains rather few nonpyramidal neurons (Valverde and Lopez-Mascaraque 1981). Accordingly, fewer GABAergic cells are found in the pallium of the goldfish than in the cortex of the lizard and the rat (our unpublished data), and the absolute content of the GABA synthesizing enzyme glutamic acid decarboxylase (GAD) is higher in the pigeon and rat (Bissoli et al. 1988) than in the frog and turtle (Bissoli and Contestabile 1988). In "primitive" mammalian species, such as the hedgehog, the neocortex (Valverde 1983) and the archicortex (Stephan and Manolescu 1980) contain fewer nonpyramidal cells than in primates. Likewise, a higher complexity of the cortex seems to be paralleled by an increased number of morphologically different types of nonpyramidal neurons (e.g. Valverde 1983); thus the morphology of GABAergic neurons is rather homogenous in the goldfish as compared to lizard or rat (Fig. 2).

Histochemistry and Immunocytochemistry

Many of the techniques which yield useful results concerning the lamination and delimitation of cortical areas in the mammalian brain, such as the staining for acetyl cholinesterase (AChE), cytochromoxidase and zinc, have been applied in studies on nonmammalian brains. However, a substantial amount of data is only available for reptiles, in which the results so far have not been unequivocal. Some authors claim the existence of close similarities between the staining pattern of enzymes and neurotransmitters whereas others stress the differences that are apparent in their material. Thus, Smeets (1988) reported a considerable variation in the monoaminergic systems. AChE staining resulted in rather different patterns even within the lizards (cf. Northcutt 1978; Perez-Clausell and Fredens 1988; Regidor and Poch 1988), in which, however, no great variation was found in the immunoreactivity for somatostatin, enkephalin and cholecystokinin (Perez-Clausell and Fredens 1988) and in the distribution of zinc (Perez-Clausell and Fredens 1988; Lopez-Garcia et al., this volume) as visualized by Timm staining.

Cholinergic (e.g. Brantley and Bass 1988; Brauth et al. 1985; Müller 1987; Contestabile et al., this volume) and GABAergic neurons (Schwerdtfeger and Lorente 1988; Müller, this volume; Contestabile et al., this volume) occur in the pallium of all teleosts and terrestrial nonmammalian species investigated so far. The same may be valid for substance P-ergic (see Veenman, this volume) and dopaminergic (Smeets et al. 1986b, 1987) neurons and for the serotonergic input to the pallium from the midbrain raphe nuclei (see Wicht and Himstedt, this volume). In all classes, glutamate/aspartate is probably the most important excitatory transmitter of pallial projection neurons (Contestabile et al., this volume) and GABA the most abundant inhibitory substance.

Similar interactions between cholinergic and GABAergic systems have been suggested for sensory cortices of birds and mammals (Müller, this volume), whereas nonmatching results have been obtained concerning the interaction of dopamine and acetylcholine in the rat and the lizard (Stoof et al. 1987).

Since the zinc-rich, and therefore highly Timm-positive, mossy fiber system is the most outstanding histochemical feature of the mammalian hippocampus (e.g. Haug 1973), Timm staining has also been used in studies dealing with the brains of submammalian species. However, the only results allowing comparisons with mammals have been found in squamate reptiles which possess a Timm-positive system with a clearly higher staining intensity than that of the other telencephalic fiber systems (Martinez-Garcia and Olucha 1988; Perez-Clausell and Fredens 1988). In the chicken, Timm staining of the hippocampus is fainter than that of several other forebrain structures (Faber and Scheich 1988). The cortex of nonsquamate reptiles and the pallium of fishes and amphibians do not contain highly Timm-positive zones (Lopez-Garcia and Perez-Clausell, personal communications).

Taken together, one gets the impression that, at present, histochemistry and immunocytochemistry can support only some rather general similarities between the pallial zones of nonmammals and mammals. It is conceivable that functional reasons may have caused a diversified development of the chemical systems; however, better insights will certainly be obtained by the study of a distinctly greater number of nonmammalian species.

Fiber Connections

The comparison of the organization of fiber connections may greatly aid the study of putative homologies between telencephalic regions. However, with the modern techniques of tracing intraaxonal, retrograde and anterograde transport, mammalian brains have mainly been investigated. Only in the most recent past have these methods also been applied to brains from species of other classes. Here, the best documented are the olfactory connections, probably because the olfactory bulb is easily accessible for experimental lesioning and injecting and does not pose a fibers-of-passage problem. Certainly, results on the course and termination mode of the secondary and tertiary olfactory fibers should yield valuable data, e.g. for discussion of the question whether the *primordium hippocampi* of fish brains is located in medial or in lateral pallial zones. The existing evidence so far has led to opposite views on this problem (Northcutt and Braford 1980; Schnitzlein 1982, Murakami et al. 1983; see also Rooney and Szabo, this volume). By contrast, there is agreement that a posterior part of the dorsal pallium is the main target of secondary olfactory projections and thus a putative homologue of the lateral pallium/lateral cortex (Northcutt and Braford 1980; Murakami et al. 1983).

A basic set of pallial afferent and efferent connections has been detected in all vertebrate classes: afferents from the olfactory bulb, septum, dorsal thalamus and brain stem, and efferents to the septum, hypothalamus and thalamus. In amphibians, reptiles and mammals, the main olfactory input reaches lateral fields (lateral pallium, lateral cortex, entorhinal cortex) which, in turn, project to medial fields (medial pallium, Scalia 1976; medial cortex, Halpern 1980; hippocampus, Hjorth-Simonsen 1972). The latter are reciprocally related to the septum and receive afferents from anterior thalamic nuclei (medial pallium, Wicht and Himstedt 1988; medial cortex, Desan, 1988; Martinez-Garcia and Lorente, this volume; hippocampus, Schwerdtfeger 1984). **Fiber connectivity hence supports the idea of a homology of medial pallium, medial cortex and hippocampus.**

Most interesting with respect to phylogenetic development of periarchicortex and neocortex is the general cortex (dorsal pallium of amphibians, dorsal cortex of reptiles, parahippocampus of birds) that apparently processes visual, auditory and other sensory information (e.g. Northcutt and Kicliter 1980; Benowitz 1980; Belekhova and Ivazov 1983). In reptiles, the dorsal cortex is composed of three subfields of which one may represent a "visual cortex" (it has been described to be the target of fibers from the lateral geniculate nucleus in turtles; Ulinski 1988), while another has been compared with mammalian subiculum (Hoogland and Vermeulen-van der Zee 1988), and entorhinal cortex (Desan 1988). The dorsal cortex hence may be considered a forerunner of some archicortical and neocortical areas, implying that it also brought forth periarchicortical areas, such as the presubicular and retrosplenial regions, which in mammals are interposed between neocortex and hippocampus.

Functional Considerations

Archicortical and periarchicortical areas have generally been considered parts of a "limbic" system to which roles in emotional and sexual behavior, orientation, visceral regulation and memory and learning have been ascribed. The capacity of learning, which has been clearly demonstrated for all phylogenetic groups of vertebrates, is apparently impeded by telencephalic lesions. However, a significant ability for memory and learning is probably related to a high development of the telencephalon as a whole; in the mammalian brain, the morphological basis of memory is apparently constituted not by one particular brain structure, but by a network of functionally associated neurons that also participate in the formation of the neural circuits and chains composing the so-called "functional systems". Such relations may explain: (1) why experimental damage to a given forebrain structure may impair a certain function, whereas other experimental approaches support participation of this structure in rather different functions and (2) the impairment of the same function by lesions of different structures.

Anatomical and physiological evidence indicates a close functional relation between the main archicortical and periarchicortical regions of the mammalian brain, namely, between the hippocampus and the entorhinal cortex (e.g. Swanson et al. 1982). One current view on the function of the hippocampus concerns its possible role in the proper classification of newly acquired information and comparison of the latter with stored information (Vinogradova 1975). The hippocampus thus may help the animal to shift its attention (cf. Stephan 1975; Schwerdtfeger 1984; Winocur 1987; Foreman and Stevens 1987) to relevant stimuli

3

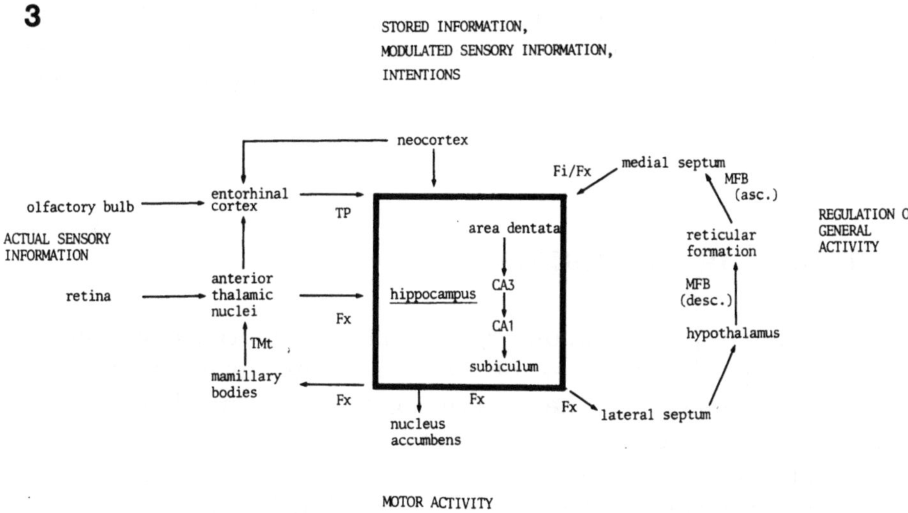

Fig. 3. Some anatomical and putative functional relations of mammalian brain structures belonging to the limbic system. *CA1, CA3,* fields CA1 and CA3 of hippocampus; *Fi,* fimbria; *Fx,* fornix; *MFB (asc.)* and *MFB (des.),* ascending and descending portions, respectively, of medial forebrain bundle; *TMt,* mamillothalamic tract; *TP,* perforant path

and to react to the latter in an appropriate manner. Hippocampus and entorhinal cortex are equipped with the necessary fiber connections to receive information on external, e.g. visual, auditory (from neocortex) and olfactory (from olfactory bulb) stimuli and on internal stimuli related to the brain stem and hypothalamus. Moreover, their efferents reach, for example, motor centers (hippocampus to nucleus accumbens; entorhinal cortex to motor cortex) and internal control centers (hypothalamus) and may thus form part of a "visceral", unconscious control system (Fig. 3).

Attention-related mechanisms that help the animals to cope with unfamiliar situations have also been described for the telencephalon of fish (Farr and Savage 1978), amphibians (Ewert and Finkenstädt 1987) and birds (Zeier 1971; Veenman, this volume). There is evidence that, within the telencephalon, the archicortex (medial pallium, hippocampus) may be crucial for the performance of that function (Schwerdtfeger 1984; Foreman and Stevens 1987; Winocur 1987), whereas the major function of the paleocortex is the relay and processing of olfactory information. The third of the principal elements of the pallium, the dorsal pallium/ dorsal cortex/neocortex, receives various types of input mainly from the thalamus and may be viewed as the highest center of sensory integration and motor control. These properties have been improved during mammalian evolution by increased capacities of storage and association, warranted by the enormous increase in the number of available neurons.

Concluding Remarks

Considering our present knowledge on the anatomical, histochemical and physiological properties of the vertebrate telencephalon, one gets the impression that the archicortex of the mammalian and bird brain, the medial/dorsomedial cortex of reptiles and the medial pallium of amphibians should be phylogenetically related. Likewise, similarities are obvious between the lateral pallium, lateral cortex and entorhinal cortex and between amphibian dorsal pallium, part of reptilian dorsal cortex and mammalian neocortex. For the different classes of fishes, no clear statement on such relations of the pallial divisions can, at present, be given. Obviously, comprehensive investigations on a larger sample of species will have to be performed. Such studies will be aided by the combined application of now available techniques. Thus, recently developed combinations of several neuroanatomical methods allow one to trace fiber connections over more than one synapse (Buhl et al. 1989; Schwerdtfeger et al. 1990) and to concomitantly identify the involved neuronal types. Notable interdisciplinary approaches are the screening of neuronopathic and behavioral effects of injected drugs (e.g. Greenberg et al. 1988) and the combination of anatomical and physiological approaches (e.g. Misgeld and Frotscher 1986; Shen and Kriegstein 1986).

Only after the evaluation of comparative studies, using the whole spectrum of available techniques, will sounder phylogenetic interpretations of the telencephalic structures of recent vertebrates be warranted.

References

Amaral DG (1978) A golgi study of cell types in the hilar region of the hippocampus of the rat. J Comp Neurol 182:851-915

Angevine JB, Sidman RL (1962) Autoradiographic study of histogenesis in the cerebral cortex of the mouse. Anat Rec 142:210

Belekhova MG, Ivazov NI (1983) Analysis of transmission of visual, somatic and audiovibrational sensory information to the hippocampal cortex of lizards. Neirofiziologiia 15:153-159

Benowitz L (1980) Functional organization of the avian telencephalon. In: Ebbesson SOE (ed) Comparative neurology of the telencephalon. Plenum, New York

Bissoli R, Contestabile A (1988) Evolution of neurotransmitter-related markers in the vertebrate telencephalon. Comparative microchemical study in discrete brain regions of a frog and a turtle. Comp Biochem Physiol 89C:241-248

Bissoli R, Battistini S, Guarnieri T, Contestabile A (1988) Regional levels of neurotransmitter markers in the pigeon telencephalon: a comparison with possibly homologous areas of the rat telencephalon. J Neurochem 50:1731-1737

Brantley RK, Bass AH (1988) Cholinergic neurons in the brain of a teleost fish *(Porichthys notatus)* located with a monoclonal antibody to choline acetyltransferase. J Comp Neurol 275:87-105

Brauth SE, Kitt CA, Price DL, Wainer BH (1985) Cholinergic neurons in the telencephalon of the reptile, *Caiman crocodilus*. Neurosci Lett 58:235-240

Buhl EH, Schwerdtfeger WK, Germroth P (1989) New anatomical approaches to reveal afferent and efferent hippocampal circuitry. In: Chan-Palay V and Köhler C (eds) The hippocampus - new vistas. Liss, New York, pp 71-83

Desan PH (1988) Organization of the cerebral cortex in turtle. In: Schwerdtfeger WK, Smeets WJAJ (eds) The forebrain of reptiles. Karger, Basel, pp 1-11

Ebbesson SOE, Voneida TJ (1969) The cytoarchitecture of the pallium in the tegu lizard (*Tupinambis nigropunctatus*). Brain Behav Evol 2:431-466

Ewert JP, Finkenstädt T (1987) Modulation of tectal functions by prosencephalic loops in amphibians. Behav Brain Sci 10:122-123

Faber H, Scheich H (1988) Characterization of zinc positive structures with Timm's silver staining in the domestic chick brain. In Elsner N, Barth FG (eds) Sense organs. Interfaces between environment and behavior. Proceedings of the 16th Göttingen Neurobiology Conference. Thieme, Stuttgart, p 174

Farr EJ, Savage GE (1978) First- and second-order conditioning in goldfish and their relation to the telencephalon. Behav Biol 22:50-59

Filimonoff IN (1947) A rational subdivision of the cerebral cortex. Arch Neurol Psychiatry 58:296-311

Foreman N, Stevens R (1987) Relationships between the superior colliculus and hippocampus: neural and behavioral considerations. Behav Brain Sci 10:101-152

Germroth P, Schwerdtfeger WK, Buhl EH (1989) GABAergic neurons in the entorhinal cortex project to the hippocampus. Brain Res 494:187-192

Greenberg N, Font E, Switzer RC III (1988) The reptilian striatum revisited: Studies on *Anolis* lizards. In: Schwerdtfeger WK, Smeets WJAJ(eds) The forebrain of reptiles. Karger, Basel, pp 162-178

Halpern M (1980) The telencephalon of snakes. In Ebbesson SOE (ed) Comparative neurology of the telencephalon. Plenum, New York, pp 257-295

Haug F-MS (1973) Heavy metals in the brain. A light microscopic study in the rat with Timm's sulphide silver method. Methodological considerations and cytological and regional staining patterns. Adv Anat Embryol Cell Biol 47:1-71

Hjorth-Simonsen A (1972) Projection of the lateral part of the entorhinal area to the hippocampus and fascia dentata. J Comp Neurol 146:219-232

Hoogland PV, Vermeulen-van der Zee E (1988) Intrinsic and extrinsic connections of the cerebral cortex of lizards. In: Schwerdtfeger WK, Smeets WJAJ (eds) The forebrain of reptiles. Karger, Basel, pp 20-29

Johnston JB (1924) Further contributions to the study of the evolution of the forebrain. V. Survey of forebrain morphology. J Comp Neurol 36:143-192

Kuhlenbeck H (1927) Vorlesungen über das Zentralnervensystem der Wirbeltiere. Fischer, Jena.

Kuhlenbeck H (1929) Die Grundbestandteile des Endhirns im Lichte der Bauplanlehre. Anat Anz 67:1-51

Kuhlenbeck H (1977) The central nervous system of vertebrates, vol. 5/I. Karger, Basel, pp 650-687

Lorente de Nó R (1933) Studies on the structure of the cerebral cortex. I. The area entorhinalis. J Psychol Neurol (Leipz) 45:381-438

Lorente de Nó R (1934) Studies on the structure of the cerebral cortex. II. Continuation of the study of the ammonic system. J Psychol Neurol (Leipz) 46:113-177

Lund JS (1984) Spiny stellate neurons. In: Peters A, Jones EG (eds) Cerebral cortex, vol I: Cellular components of the cerebral cortex. Plenum, New York, pp 255-308

Lund JS, Boothe R (1975) Interlaminar connections and pyramidal neuron organization in the visual cortex, area 17, of the macaque monkey. J Comp Neurol 159:305-334

Martinez-Garcia F, Olucha FE (1988) Afferent projections to the Timm positive cortical areas of the telencephalon of lizards. In: Schwerdtfeger WK, Smeets WJAJ (eds) The forebrain of reptiles. Karger, Basel, pp 30-40

McVicar BA, Dudek FE (1980) Local synaptic circuits in rat hippocampus: interactions between pyramidal cells. Brain Res 184:220-223

Misgeld U, Frotscher M (1986) Postsynaptic GABAergic inhibition of non-pyramidal neurons in the guinea-pig hippocampus. Neuroscience 19:193-206

Müller CM (1987) Differential effects of acetylcholine in the chicken auditory neostriatum and hyperstriatum ventrale - studies in vivo and in vitro. J Comp Physiol [A] 161:857-866

Murakami T, Morita Y, Ito H (1983) Extrinsic and intrinsic fiber connections of the telencephalon in a teleost, *Sebastiscus marmoratus*. J Comp Neurol 216:115-131

Northcutt RG (1978) Forebrain and midbrain organization in lizards and its evolutionary significance. In: Greenberg N, McLean PD (eds) The behavior and neurology of lizards. Rockville, NIMH, pp 11-64

Northcutt RG, Braford MR (1980) New observations on the organization and evolution of the telencephalon in ray finned fishes. In: Ebbesson SOE (ed) Comparative neurology of the telencephalon. Plenum, New York, pp 203-255

Northcutt RG, Kicliter E (1980) Organization of the amphibian telencephalon. In: Ebbesson SOE (ed) Comparative neurology of the telencephalon. Plenum, New York, pp 203-255

Pérez-Clausell J, Fredens K (1988) Chemoarchitectonics in the telencephalon of the lizard *Podarcis hispanica*. Distribution of somatostatin, enkephalin, cholecystokinin and zinc. In Schwerdtfeger WK, Smeets WJAJ (eds) The forebrain of reptiles. Karger, Basel, pp 85-96

Peters A, Jones EG (1984) Classification of cortical neurons. In Peters A, Jones EG (eds) Cerebral cortex, vol I: cellular components of the cerebral cortex. Plenum, New York, pp 107-122

Regidor J, Poch L (1988) Histochemical analysis of the lizard cortex: an acetylcholinesterase, cytochrome oxidase and NADPH-diaphorase study. In Schwerdtfeger WK, Smeets WJAJ (eds) The forebrain of reptiles. Karger, Basel, pp 77-84

Rehkämper G, Haase E, Frahm HD (1988) Allometric comparison of brain weight and brain structure volumes in different breeds of the domestic pigeon, *Columbia livia f.d.* (Fantails, Homing Pigeons, Strassers). Brain Behav Evol 31:141-149

Scalia F (1976) Structure of the olfactory and accessory olfactory systems. In: Llinas R, Precht W (eds) Frog neurobiology. Springer, New York, Berlin, Heidelberg, pp 213-233

Schnitzlein HN (1982) Telencephalon of fishes. In Crosby EC, Schitzlein HN (eds) Comparative correlative neuroanatomy of the vertebrate telencephalon. Mac Millan, New York, pp 3-160

Schwerdtfeger WK (1984) Structure and fiber connections of the hippocampus. Adv Anat Embyrol Cell Biol 83:1-74

Schwerdtfeger WK, Buhl EH (1986) Various types of non-pyramidal hippocampal neurons project to the septum and contralateral hippocampus. Brain Res 386:146-154

Schwerdtfeger WK, Lorente MJ (1988) Laminar distribution and morphology of gamma-aminobutyric acid (GABA)-immunoreactive neurons in the medial and dorsomedial areas of the cerebral cortex of the lizard *Podarcis hispanica*. J Comp Neurol 278:473-485

Schwerdtfeger WK, Buhl EH, Germroth P (1990) Disynaptic olfactory input to the hippocampus mediated by stellate cells in the entorhinal cortex. J Comp Neurol (in press)

Shen JM, Kriegstein AR (1986) Turtle hippocampal cortex contains distinct cell types, burst firing neurons, and an epileptogenic subfield. J Neurophysiol 56:1626-1649

Smeets WJAJ, Hoogland PV, Lohman AHM (1986a) A forebrain atlas of the lizard *Gekko gecko*. J Comp Neurol 254:1-19

Smeets WJAJ, Hoogland PV, Voorn P (1986b) The distribution of dopamine immunoreactivity in the forebrain and midbrain of the lizard *Gekko gecko*. J Comp Neurol 253:46-60

Smeets WJAJ, Jonker AJ, Hoogland PV (1987) Distribution of dopamine in the forebrain and midbrain of the red-eared turtle *Pseudemys scripta elegans*, reinvestigated using antibodies against dopamine. Brain Behav Evol 30:121-142

Smeets WJAJ (1988) The monoaminergic systems of reptiles investigated with specific antibodies against serotonin, dopamine, and noradrenalin. In Schwerdtfeger WK, Smeets WJAJ (eds) The forebrain of reptiles. Karger, Basel, pp 97-109

Smith GE (1901) Notes upon the natural subdivision of the cerebral hemisphere. J Anat 35:431-454

Smith GE (1903) On the morphology of the cerebral commissures in the Vertebrata, with special reference to an aberrant commissure found in the forebrain of certain reptiles. Trans Linn Soc (Lond), 2nd Ser, Zool 8:455-500

Somogyi P (1979) An interneuron making synapses specifically on the axon initial segment (AIS) of pyramidal cells in the cerebral cortex. J Physiol (Lond) 296:18P

Starck D (1978) Vergleichende Anatomie der Wirbeltiere, vol 1. Springer, Berlin, Heidelberg, New York

Stephan H (1975) Allocortex. In Bargmann W (ed) Handbuch der mikroskopischen Anatomie des Menschen, part 4, vol 9. Springer, Berlin, Heidelberg, New York

Stephan H, Manolescu J (1980) Comparative investigations on hippocampus in insectivores and primates. Z Mikrosk-Anat Forsch 94:1025-1050

Stephan H, Baron G, Frahm HD (1988) Comparative size of brains and brain components. In: Steklis HD, Erwin J (eds) Comparative primate biology, vol.4: neurosciences. Liss, New York, pp 1-38

Stoof JC, Russchen FT, Verheijden PFHM, Hoogland PV (1987) A comparative study of dopamine-acetylcholine interaction in telencephalic structures of the rat and of a reptile, the lizard *Gekko gekko*. Brain Res 404:273-291

Swanson LW, Teyler TJ, Thompson RF (1982) Hippocampal long-term potentiation: mechanisms and implications for memory. Neurosci Res Program Bull 20:613-764

Ulinski PS (1977) Intrinsic organization of snake medial cortex: an electron microscopic and Golgi study. J Morphol 152:247-280

Ulinski PS (1988) Functional architecture of turtle visual cortex. In Schwerdtfeger WK, Smeets WJAJ (eds) The forebrain of reptiles. Karger, Basel, pp 151-161

Valverde F (1983) A comparative approach to neocortical organization based on the study of the brain of the hedgehog (*Erinaceus europeus*). In Grisolia S, Guerri C, Samson F, Norton S, Reinoso-Suarez F (eds) Ramon y Cajal's contribution to the neurosciences, Elsevier, Amsterdam, pp 149-170

Valverde F, Lopez-Mascaraque L (1981) Neocortical endeavor: basic neuronal organization in the cortex of hedgehog. In: 11th International Congress of Anatomy: glial and neuronal cell biology. Liss, New York, pp 281-290

Vinogradova OS (1975) Functional organization of the limbic system in the process of registration of information: facts and hypotheses. In: Isaacson RL, Pribram KH (eds) The hippocampus, vol II. Neurophysiology and behavior. Plenum, New York, pp 3-70

Wicht H, Himstedt W (1988) Topologic and connectional analysis of the dorsal thalamus of *Triturus alpestris* (Amphibia, Urodela, Salamandridae) J Comp Neurol 267:545-561.

Winocur G (1987) The hippocampus and attention. Behav Brain Sci 10:132-133

Zeier H (1971) Archistriatal lesions and response inhibition in the pigeon. Brain Res 31:327-339

Prefrontal Cortex: The Name and the Thing

I. Divac[1] and R.G.E. Öberg[2]
[1]Institute of Neurophysiology, University of Copenhagen, Denmark
[2]Department of Neurology, The National Hospital, Copenhagen, Denmark

In recent decades, methodological advances have led to considerable modifications of many concepts in comparative neuroanatomy. An example is the revolutionary change in the understanding of the organization of the bird brain: the older interpretation (Crosby et al. 1967) differs radically from the presently accepted one (Karten 1969). These conceptual changes, however, have not affected the terminology proposed by the old school. Thus, the "neostriatum" in birds compares to the cerebral cortex in mammals rather than to the mammalian neostriatum. This and similar mismatches remind us of the old dictum that "words are not things". The purpose of the present essay is to show the relevance of this dictum to comparative research of the prefrontal cortex.

Owen (1868) coined the term prefrontal to denote the cortex rostral to the frontal sulcus, which he claimed he could identify in different gyrencephalic brains (see Divac 1988). In the course of time, several other terms have been introduced to refer to the anterior part of the cerebral cortex. The most frequently encountered are: frontal lobes, anterior frontal cortex, frontal association cortex, silent cortex, orbitofrontal cortex and frontal granular cortex. In principle, each of these terms is defined differently and refers to a cortical area which more or less overlaps an area described by a different term. Yet the literature shows that these terms are often used as synonyms both in studies on a single species and in comparative research. The terms with topographic connotation (frontal lobes, anterior frontal cortex) are imprecise for comparisons of lissencephalic brains and gyrencephalic brains which have different gyral patterns. These terms may include the premotor and motor areas. The functional term "association" implies an as yet unattained understanding of function, and the silence of a frontal area is relative (see below). Furthermore, at present, the functional criteria do not clearly delineate these areas. The term orbitofrontal was used by Rose and Woolsey (1948) to describe the cortical target of the thalamic mediodorsal nucleus in non-primates.

Experimental Brain Research Series 19

This term seems unnecessary because there is no obvious reason to distinguish the so defined area in primates and non-primates. In addition, this term is used with the topographical term "orbital frontal", used for a subdivision of the prefrontal area in primates. Strictly speaking, the term "frontal granular cortex" applies only to one part of the frontal area and only to primates; a conspicuous fourth granular layer is absent in the orbital frontal cortex (of primates) and the entire orbitofrontal cortex (of non-primates).

The term prefrontal has all but lost its topographical connotation. It has been relatively precisely redefined, and therefore it seems most suitable for comparative purposes in spite of its being linguistically awkward.

Unfortunately, the term prefrontal is not singularly defined. Most often, the term denotes the cortical area innervated by the thalamic mediodorsal nucleus (Rose and Woolsey 1948), but sometimes topographic and architectonic (e.g. Brodmann 1912) definitions are used instead. Even seasoned contemporary authors tend to assume that the different definitions are interchangeable. For example, Fuster (1989) defines the prefrontal cortex on the basis of hodology, cytoarchitecture and topography in different sections of his monograph.

Ideally and ultimately, anatomical entities should be defined on the basis of exhaustive knowledge of their phylogenetic and ontogenetic (embryogenetic) properties, as well as chemistry, structure and function in adulthood. In the absence of such knowledge, however, we must rely on practical and well-defined terminology. In general, investigators who work with a single species may choose any unequivocally defined part of the brain as the entity worthy of study. Attempts to study different species tend to constrict this freedom. The larger the number of species to be compared, and the more distant the relationships among them, the more difficult it is to establish criteria for inclusive definitions.

In the rhesus monkey, we appear to be able to define interchangeably the prefrontal cortex as the cortex of the frontal lobe which is situated in the front of the arcuate sulcus dorsolaterally and the *tuberculum olfactorium* ventrally. This cortical area is irresponsive to electrical stimulation (Fulton 1949), has a conspicuous inner granular layer (Brodmann 1912) and is innervated by the thalamic mediodorsal nucleus (Akert 1964). A closer look, however, reveals problems: the topographic borders between the prefrontal cortex and the premotor areas above and below the arcuate sulcus, as well as on the mesial surface are not clear; electrical stimulation of the orbital cortex does induce measurable changes in breathing and in some autonomic functions (Delgado and Livingston 1948; Livingston et al. 1948); in addition, the fourth layer of the orbital prefrontal area is indistinct (Akert 1964). Notice that the first three criteria include the stipulation that the prefrontal area is in the frontal lobe, i.e. they include an element of topography.

The problems encountered in the comparative study of the prefrontal cortex may be illustrated by trying to apply the above characterizations to the rat. As mentioned above, topography is of little help in a lissencephalic brain. Classical cytoarchitectonics ruled that "lower" species, such as the rabbit, have a very small prefrontal area (occupying only 2.2% of the total cortical surface), and that even more primitive species, such as the hedgehog, have none (Brodmann 1912). In the rat and mouse, a small area, located dorsally on the frontal pole, was, and sometimes still is claimed to have an identifiable fourth granular layer (Rose 1929; Van Eden and Uylings 1985), but other investigators disagree (Zilles et al. 1980). Study of cytoarchitecture suggested to Krieg (1946) that the prefrontal area occupied the dorsal surface of the frontal lobe in rats. This area was assumed to be comparable to the prefrontal cortex of other mammals until 1969 when Leonard demonstrated that the mediodorsal cortex projects to separate fields in the mesial and the suprarhinal frontal cortex, avoiding the dorsal pole (Krettek and Price 1977; Divac et al. 1978b; Groenewegen 1988). Thus, in the rat, the topographic and cytoarchitectonic criteria, and the hodological criteria, do not coincide.

For decades, neurobehavioral studies on the rat were guided by the assumption that the frontal pole is comparable to the prefrontal area of primates (e.g. Gross et al. 1965). Following the hodological data of Leonard (1969), we ablated either the mesial or the dorso-polar cortical area in different groups of rats and looked for neurobehavioral impairments characteristic for prefrontal dysfunction in primates and carnivores (Konorski and Lawicka 1964; Rosvold and Szwarcbart 1964). Only the group with mesial lesions showed characteristic impairments in spatial reversals and delayed alternation (Divac 1971; Srebro and Divac 1972; Wikmark et al. 1973). These results have since been confirmed in numerous other experiments on rats (Kolb 1984). Similar studies on other species, e.g. tree shrews (Divac et al. 1978a; Passingham 1978; Divac and Passingham 1980) and guinea pigs (Markowitsch and Riess 1981), also demonstrated that ablation of the cortical target of the mediodorsal nucleus impairs delayed response behavior. These data suggest that the cortical area, which receives projections from the mediodorsal nucleus, regardless of its topographic position and cytoarchitectonic appearance, mediates the same function in these species. This presumed common function is found in species with as different brains, ecological niches and behavioral repertoires as those of rats, cats, dogs and monkeys. The prefrontal cortex in different species shows further functional similarities, summarized by Kolb (1984).

In addition to behavioral function (inferred mainly from lesion effects), certain neurochemical properties are in register with the projections from the thalamic mediodorsal nucleus. The target area of the mediodorsal nucleus receives a denser dopaminergic innervation than any other part of the cerebral cortex in mammals of different lines of descent (Björklund et al. 1978; Divac et al. 1978a; Brown et al. 1979). Hence, a dense dopamine-innervation may guide the search for the prefrontal cortex. This approach seems to have revealed tissue equivalent to prefrontal cortex in pigeons whose thalamus, according to Le Gros Clark (1932), does not contain a distinct mediodorsal nucleus (Mogensen and Divac 1982; Divac et al. 1985; Björklund and Mogensen 1985). This structure lies in the postero-dorso-lateral neostriatum. Its localization in the caudal end of the forebrain matches the rostral localization of formations involved in vision.

At present it is not known whether reptiles have an equivalent of the mammalian prefrontal cortex as well. Studies which describe the distribution of dopamine-containing fibers offer some tantalizing hints: In a lizard *Gecko gecko*, an accumulation of dopamine-containing fibers is seen in the postero-dorso-lateral portion of the dorsal ventricular ridge, resembling the position of the hypothetical equivalent of the mammalian prefrontal cortex in pigeons (Smmets et al. 1986, Fig. 2E,F). Such an accumulation is not conspicuous in a snake, *Python regius* (Smeets 1988), or a turtle, *Pseudemys scripta elegans* (Smeets et al. 1987). Dopamine-containing fibers are, however, seen in the dorsal ventricular ridge of both of these species. It is possible that in birds and reptiles the topography of the "prefrontal cortex" varies as it does in mammals (e.g. Divac et al. 1978a).

The hodological definition of the prefrontal cortex is not without problems. Some neurons in the mediodorsal nucleus project to parietal and temporal cortical areas, and other thalamic nuclei overlappingly project to the prefrontal area (e.g. Kievit and Kuypers 1977; Divac 1979). These complications were reviewed by Markowitsch and Pritzel (1979) who concluded that thalamic innervation provided an inadequate basis for definition, and who proposed a research program aimed at a polythetic definition (Soral 1974). Current knowledge does not allow a polythetic definition, and thus, dismissal of the efferents from the mediodorsal nucleus is not warranted as the defining feature. The number of neurons in this nucleus which project to the frontal lobe in mammals is more than one order of magnitude larger than the number projecting elsewhere in the brain. The definition of the prefrontal cortex could therefore be modified to: the cortical target of the *densest* projection from the mediodorsal nucleus. This formulation resembles Rose and Woolsey's (1948) definition of the orbitofrontal cortex as the area which receives "essential" projections from the mediodorsal nucleus. It should be noted that all neurons on the conventional territory of this nucleus need not be hodologically and chemically like the rest; rather, it is conceivable that some perikarya belonging to intralaminar neurons become displaced into the mediodorsal cortex but keep their connections to different cortical areas unchanged.

The overlap of projections from the mediodorsal nucleus with projections of other thalamic nuclei, such as the anteromedial and medioventral (Divac et al. 1978b), should be related to the observation of Domesick (1972), that the mediodorsal nucleus projects to the middle layers, like other specific thalamic nuclei, while the anteromedial nucleus projects to the layers I and VI. This may mean that the latter projections, as well as those from the medioventral (axial) nucleus (Divac et al. 1978b; Herkenham 1979), are not specific and therefore do not define cortical areas.

In our opinion, the most serious problem with the present hodological definition of the prefrontal cortex is that it presupposes that the mediodorsal nucleus can be reliably identified. Unfortunately, we rely presently on topography (position in the thalamus and the relation with the *lamina medullaris interna*, wherever the latter is identifiable), and low levels of chemical markers such as acetylcholinesterase, cytochrome oxidase, most peptides and membrane receptors (review in Regidor and Divac 1987). Afferents to the mediodorsal nucleus may become an important guide to its positive identification (e.g. Sapawi and Divac 1978; Russchen et al. 1987; Cornwall and Phillipson 1988; Groenewegen 1988).

In conclusion, we currently accept the definition of the prefrontal cortex as the cortical target of the (densest) projections of the thalamic mediodorsal nucleus for the following reasons: (1) the projections from the mediodorsal nucleus correlate highly with functional and chemical properties; (2) this definition is consistent with identification of non-prefrontal cortical areas; (3) it denotes a comparable area in all mammals. The definition was formally proposed by Rose and Woolsey (1948), it was implicit in the initial observation of thalamo-cortical relations by Monakov (1895) and it can be considered a special case of the general proposal to define cortical areas by their afferents from specific thalamic nuclei (Bonin 1944; McCulloch 1944).

Acknowledgements. The research on the prefrontal cortex in I. Divac's laboratory has been supported by the Danish Medical Research Council, Fonden of 1870 and Ib Henriksens Fond. Jesper Morgensen has had an important role in this research. Margit Løvgreen provided secretarial help.

References

Akert K (1964) Comparative anatomy of frontal cortex and thalamo-cortical connections. In: Warren JM, Akert K (eds) Frontal granular cortex and behavior. McGraw-Hill, New York, pp 372-396

Björklund A, Divac I, Lindvall O (1978) Regional distribution of catecholamines in monkey cerebral cortex. Evidence for a dopaminergic innervation of the primate prefrontal cortex. Neurosci Lett 7:115-119

Bonin G v (1944) The architecture of the precentral motor cortex. In Bucy PC (ed) The precentral motor cortex. Ill Monogr Med Sci, Urbana Ill., pp 7-82

Brodmann K (1912) Neue Ergebnisse über die vergleichende histologische Lokalisation der Grosshirnrinde mit besonderer Berücksichtigung des Stirnhirns. Anat Anz (Suppl) 41:157-216

Brown RM, Crane AB, Goldman PS (1979) Regional distribution of monoamines in the cerebral cortex and subcortical structures of the rhesus monkey: concentrations and in vivo synthesis rates. Brain Res 168:133-150

Cornwall J, Phillipson OT (1988) Afferent projections to the dorsal thalamus of the rat as shown by retrograde lectin transport. I. The mediodorsal nucleus. Neuroscience 24:1035-1049

Crosby EC, De Jonge BR, Schneider RC (1967) Evidence for some of the trends on the phylogenetic developments of the vertebrate telencephalon. In Hassler R, Stephan H (eds) Evolution of the forebrain. Plenum, New York, pp 117-135

Delgado JMR, Livingston RB (1948) Some respiratory, vascular and thermal responses to stimulation of orbital surface of frontal lobe. J Neurophysiol 11:39-55

Divac I (1971) Frontal lobe system and spatial reversal in the rat. Neuropsychologia 9:175-183

Divac I (1979) Patterns of subcortico-cortical projections as revealed by somatopetal horseradish peroxidase tracing. Neuroscience 4:455-461

Divac I (1988) A note on the history of the term "prefrontal". IBRO News 16:2

Divac I, Mogensen J (1985) The prefrontal "cortex" in the pigeon. Catecholamine histofluorescence. Neuroscience 15:677-682

Divac I, Passingham RE (1980) Connections of the mediodorsal nucleus of the thalamus in the tree shrew. II. Efferent connections. Neurosci Lett 19:21-26

Divac I, Björklund A, Lindvall O, Passingham RE (1978a) Converging projections from the mediodorsal thalamic nucleus and mesencephalic dopaminergic neurons to the neocortex in three species. J Comp Neurol 180:59-72

Divac I, Kosmal A, Björklund A, Lindvall O (1978b) Subcortical projections to the prefrontal cortex in the rat as revealed by the horseradish peroxidase technique. Neuroscience 3:785-796

Divac I, Mogensen J, Björklund A (1985) The prefrontal "cortex" in the pigeon. Biochemical evidence. Brain Res 332:365-386

Domesick VB (1972) Thalamic relationships of the medial cortex in the rat. Brain Behav Evol 6:457-583

Fulton JF (1949) Physiology of the nervous system, 3rd edn., Oxford University Press, Oxford

Fuster JM (1989) The prefrontal cortex 2nd edn., Raven, New York

Groenewegen HJ (1988) Organization of the afferent connections of the mediodorsal thalamic nucleus in the rat, related to the mediodorsal-prefrontal topography. Neuroscience 24:379-431

Gross CG, Chorover SL, Cohen SM (1965) Caudate, cortical, hippocampal and dorsal thalamic lesions in rats: alternation and Hebb-Williams maze performance. Neuropsychologia 3:53-68

Herkenham M (1979) The afferent and efferent connections of the ventromedial thalamic nucleus in the rat. J Comp Neurol 184:487-518

Karten HJ (1969) The organization of the avian telencephalon and some speculations on the phylogeny of the amniote telencephalon. Ann NY Acad Sci 167:164-180

Kievit J, Kuypers HGJM (1977) Organization of the thalamo-cortical connections to the frontal lobe in the rhesus monkey. Exp Brain Res 29:299-322

Kolb B (1984) Functions of the frontal cortex of the rat: A comparative review. Brain Res Rev 8:65-98

Konorski J, Lawicka W (1964) Analysis of errors by prefrontal animals on the delayed-response test. In: Warren JM, Akert K (eds) The frontal granular cortex and behavior. McGraw-Hill, NY, pp. 271-294

Krettek JE, Price JL (1977) The cortical projections of the mediodorsal nucleus and adjacent thalamic nuclei in the rat. J Comp Neurol 171:157-192

Krieg WJ (1946) Connections of the cerebral cortex. I. The albino rat. A. Topography of the cortical areas. J Comp Neurol 84:221-275

Le Gros Clark WE (1932) The structure and connections of the thalamus. Brain 55:406-470

Leonard CM (1969) The prefrontal cortex of the rat. I. Cortical projections of the mediodorsal nucleus. II. Efferent connections. Brain Res 12:321-343

Livingston RB, Fulton JF, Delgado JMR, Sachs E, Brendler SJ, Davis GD (1948) Stimulation and regional ablation of orbital surface of frontal lobe. Res Publ Assoc Nerv Ment Dis 27:405-420

Markowitsch HJ, Pritzel M (1979) The prefrontal cortex: projection area of the thalamic mediodorsal nucleus? Physiol Psychol 7:1-6

Markowitsch HJ, Riess R (1981) Delayed-alternation performance after selective lesions of the medial and sulcal prefrontal cortex of the guinea pig. Brain Behav Evol 18:96-104

McCulloch WS (1944) The functional organization of the cerebral cortex. Physiol Rev 3:390-407

Mogensen J, Divac I (1982) The prefrontal "cortex" in the pigeon. Behavioral evidence. Brain Behav Evol 21:60-66

Monakov CV (1895) Experimentelle und pathologisch-anatomische Untersuchungen über die Haubenregion, den Sehhügel und die *Regio subthalamica*, nebst Beiträgen zur Kenntnis früh erworbener Gross- und Kleinhirndefekte. Arch Psychiatr Nervenkrankh 27:1-128

Owen R (1868) On the anatomy of vertebrates, vol. 3, Mammals. Longmans, Green, London

Passingham RE (1978) The functions of prefrontal cortex in the tree shrew (*Tupaia belangeri*). Brain Res 145:147-152

Regidor J, Divac I (1987) Architectonics of the thalamus in echidna (*Tachyglossus aculeatus*): search for the mediodorsal nucleus. Brain Behav Evol 30:328-341

Rose JE, Woolsey CN (1948) The orbitofrontal cortex and its connections with the mediodorsal nucleus in rabbit, sheep and cat. Res Publ Assoc Res Nerv Ment Dis 27:210-232

Rose M (1929) Cytoarchitektonischer Atlas der Grosshirnrinde des Menschen. J Psychol Neurol 40:1-51

Rosvold HE, Szwarcbart MK (1964) Neural structures involved in delayed response performance. In: Warren JM, Akert K (eds) The frontal granular cortex and behavior. McGraw-Hill, New York, pp 1-15

Russchen FT, Amaral DG, Price JL (1987) The afferent input to the magnocellular division of the mediodorsal thalamic nucleus in the monkey, *Macaca fascicularis*. J Comp Neurol 256:175-210

Sapawi RR, Divac I (1978) Connections of the mediodorsal nucleus of the thalamus in the tree shrew. I. Afferent connections. Neurosci Lett 7:183-189

Sokal R (1974) Classification: purposes, principles, progress, prospects. Science 185:1115-1123

Smeets WJAJ (1988) The distribution of dopamine immunoreactivity in the forebrain and midbrain of the snake *Python regius*: a study with antibodies against dopamine. J Comp Neurol 271:115-129

Smeets WJAJ, Hoogland PV, Voorn P (1986) The distribution of dopamine immunoreactivity in the forebrain and midbrain of the lizard *Gecko gecko*: an immunohistochemical study with antibodies against dopamine. J Comp Neurol 253:46-60

Smeets WJAJ, Jonker AJ, Hoogland PV (1987) Distribution of dopamine in the forebrain and midbrain of the red-eared turtle *Pseudemys scripta elegans*, reinvestigated using antibodies against dopamine. Brain Behav Evol 30:121-142

Srebro B, Divac I (1972) Successive position reversals in rats with septal and/or fronto-polar lesions. Physiol Behav 9:269-272

Van Eden CG, Uylings HBM (1985) Cytoarchitectonic development of the prefrontal cortex in the rat. J Comp Neurol 241:253-267

Wikmark RGE, Divac I, Weiss R (1973) Retention of spatial delayed alternation in rats with lesions in the frontal lobes. Implications for a comparative neuropsychology of the prefrontal system. Brain Behav Evol 8:329-339

Zilles K, Zilles B, Schleicher A (1980) A quantitative approach to cytoarchitectonics. VI. The areal pattern of the cortex of the albino rat. Anat Embryol (Berl) 159:335-360

Subject Index

MIX
Papier aus verantwortungsvollen Quellen
Paper from responsible sources
FSC® C105338

If you have any concerns about our products,
you can contact us on
ProductSafety@springernature.com

In case Publisher is established outside the EU,
the EU authorized representative is:
Springer Nature Customer Service Center GmbH
Europaplatz 3, 69115 Heidelberg, Germany

Printed by Libri Plureos GmbH
in Hamburg, Germany